The Golden Age of
Cigarette Lighters

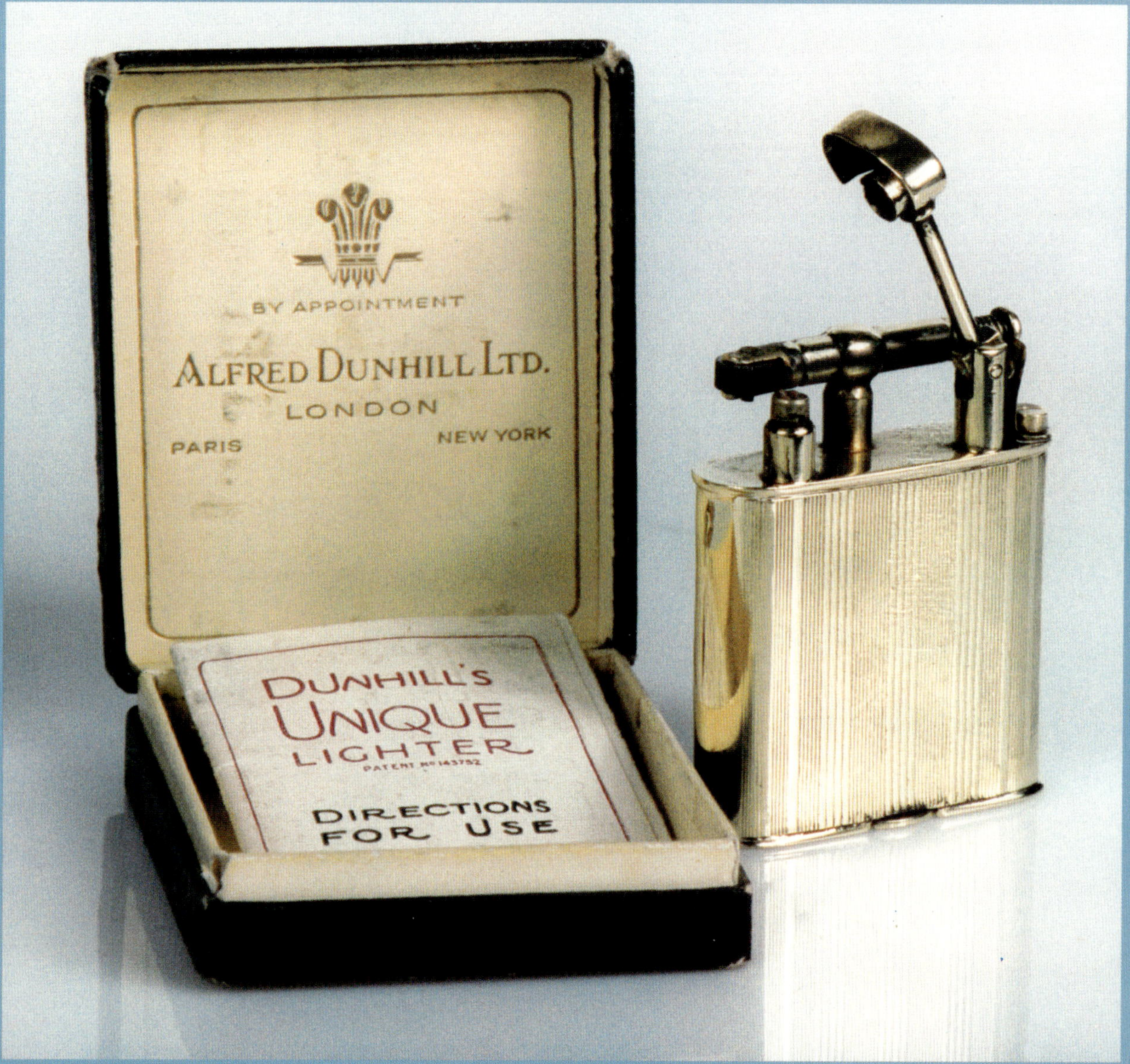

Ira Pilossof and Stuart Schneider

4880 Lower Valley Road, Atglen, PA 19310 USA

Published by Schiffer Publishing Ltd.
4880 Lower Valley Road
Atglen, PA 19310
Phone: (610) 593-1777; Fax: (610) 593-2002
E-mail: Info@schifferbooks.com
Please visit our web site catalog at **www.schifferbooks.com**
We are always looking for people to write books on new and related subjects. If you have an idea for a book, please contact us at the above address.

This book may be purchased from the publisher.
Include $3.95 for shipping.
Please try your bookstore first.
You may write for a free catalog.

In Europe, Schiffer books are distributed by
Bushwood Books
6 Marksbury Avenue
Kew Gardens
Surrey TW9 4JF England
Phone: 44 (0) 20 8392 8585
Fax: 44 (0) 20 8392 9876
E-mail: Bushwd@aol.com
Free postage in the UK. Europe: air mail at cost.

Copyright © 2004 by Ira Pilossof and Stuart Schneider
Library of Congress Control Number: 2003109064

All rights reserved. No part of this work may be reproduced or used in any form or by any means—graphic, electronic, or mechanical, including photocopying or information storage and retrieval systems—without written permission from the publisher.

The scanning, uploading and distribution of this book or any part thereof via the Internet or via any other means without the permission of the publisher is illegal and punishable by law. Please purchase only authorized editions and do not participate in or encourage the electronic piracy of copyrighted materials.

"Schiffer," "Schiffer Publishing Ltd. & Design," and the "Design of pen and ink well" are registered trademarks of Schiffer Publishing Ltd.

Cover and book designed by Bruce Waters
Type set in Americana Bold heading font/text font Humanist 521

ISBN: 0-7643-1936-1
Printed in China
1 2 3 4

Contents

Chapter One: History of Lighter Development 6

Chapter Two: Collecting Lighters 8

Chapter Three: Lighter Manufacturers 15

Chapter Four: Lighters by Country 34

    United States 34

    France 111

    Great Britain 129

    Germany 161

    Austria 168

    Italy 174

    Switzerland 175

    Japan 179

    Occupied Japan 186

    Miscellaneous 187

Glossary of Terms and Abbreviations 189

Bibliography 192

Resources 192

Preface and Acknowledgments

Being asked to do a second book on lighters was a true thrill for me. Although it is an endeavor that requires much time, when involved with a subject that you love and feel passionate about, it does not feel like work at all. Working with Stuart was always enjoyable and a true learning experience. He has a wealth of knowledge when it comes to antiques and collectibles and he loves to share his passion and experiences. We were extremely lucky to have access to some of the finest lighter collections in the United States for our book, and were greeted with open arms by fellow collectors. We also sought help and advice from some of the most widely respected names in the lighter community and are so grateful for their assistance and time. We have tried to include many of the important lighter manufacturers from the world over, providing as much detail and information about them that we could obtain. Without these knowledgeable collectors, our efforts would have been in vain. We were given access to amazing collections and if the collectors were not friends of ours before we began this endeavor, they surely are cherished friends now.

A very special thank you is extended from Stuart and myself to David Golden and Guy Nishida. David and his lovely wife, Elizabeth, made us feel completely welcome in their home during our two trips to Atlanta to photograph David's magnificent collection. David also contributed an enormous amount of his time working with us afterwards on descriptions and many other details of our book, assisting us in any way he could. His love and dedication to the hobby and his fine mind for details have enhanced our book much more than we had hoped for and we thank him so much for that.

Guy Nishida, a true friend, has been instrumental in the development of our book. We thank him for all that he has done, and are forever grateful for his time and efforts during this period. He has flown East, answered our many phone calls and e-mails, and has always been willing to share with us from his vast knowledge of lighter collecting. His love and passion for the hobby was refreshing, and inspired Stuart and I to always push on by adding, editing, and improving our material. We thank you Guy, for your time and help, and especially for your true friendship.

We have been meticulous in providing photos of the highest quality, allowing our readers to appreciate the beauty and craftsmanship of the pieces we have included in the book. Wherever possible we have tried to point out unusual characteristics of a particular lighter which might not be so apparent upon first glance. Whether it is a fabulous mechanism, a beautiful glass enamel scene, or a true Art Deco piece, there is sure to be something for everyone in this book. The 1930s and 1940s truly were a "Golden Age" in many respects, and hopefully that is reflected in the beauty and magnificence of the lighters we have chosen.

I must thank and give a big hug to my wife Susan who has been so supportive and understanding during these past two years while we worked on the book. My daughters Valerie and Julie, thank you too for being so understanding when I was not around and realizing that Dad was doing something that was very important to him. I love you all.

Sincere thank you's go out to the following people who were so important to this book: Sonny Anderson, Bob Brockmann (Zip Street, rbrockma@prodigy.net), Richard Ball, Debbie Bowman & Larry Marshall (PLPG Lighter Club, plpg1@aol.com), Alexandre Clappier, Urban & Chris Cummings, (Bird Dog Books, PO Box 1482 Palo Alto, CA 94302), John Elster, Ron Eyerkuss, Jerry Falduto, Ed Fingerman, Takashi Fujii, David Golden, Norman Israel, Hiroshi Kito, Simon M. Lytton (Engraver, England, Phone 01442 255542), Claudio Mazzi (info@zippobymazzi.it), Guy & Kathy Nishida (btrade@attbi.com), Ed Pinedo, Jonathan Rich, Maya Rubin, Judith Sanders (OTLS Lighter Club, otls@peoplescom.net), Steve Sommers, Larry Tolkin, Carlo Valerio, Richard Weinstein (ARS, info@antiquelighter.com), Kazuo Yasuge, Linda Meabon (Zippo Mfg. Co. LMeabon@zippo.com).

Foreword
What I Learned From This Project
BY DAVID GOLDEN

My "assignment" was to assist, to a limited extent, in the selection of lighters from my collection for use in the book. More significantly, I spent hours peering at lighters through a loupe and, on occasion, a digital microscope. I squinted and fiddled, trying to read proof marks, hallmarks, maker's marks, gold and silver content, etc. Then I scoured my resources, books, and records I kept as I obtained the lighters. There were the inevitable calls and emails to dealers and fellow collectors asking often insipid questions. Every lighter was measured and all the information was recorded on a spreadsheet.

Though sometimes the endeavor was tedious, it was ultimately a rewarding journey. As I took each lighter in hand, it brought back many fond memories of the people and circumstances behind its acquisition. Many were from collections I bought in whole or in part, ranging from a two-lighter collection to over one hundred pieces in a singe transaction. Every seller was different. I think that some of these sellers were happy to see their lighters wind up in the hands of an inveterate collector who cares for and enjoys the results of their endeavors. Others were perhaps a bit resentful at letting me take advantage of their years of searching, haggling, and buying individual lighters, but that is the nature of the lighter world. We come from an amazing variety of backgrounds, lifestyles, attitudes, and values.

I also learned far more information about each lighter in my collection and about the evolution of lighters on the whole. The subtle differences between a number of seemingly identical watch lighters became apparent. I began to notice slight variances in measurements from one to the other and makers' marks I never noticed before. I saw small steps in the improvement of a particular model and discovered why some lighters were made in such limited quantities (for example it did not fit well into the hand, the mechanism was not reliable, or the lighter was too fragile). Oddly, some of the most beautiful pieces are designs that just did not work well.

I found my mind roaming, wondering about who owned these lighters. How did they decide just what they wanted, as many of them clearly were special orders. Who figured out how to save a faulty design with perhaps just one minor adjustment?

I have collected lighters for many years, combining business trips with lighter hunting. During my few spare hours I would hunt down individual pieces. Lighters, being small in most instances, had the advantage of having many aspects that attracted me, such as quality, ingenuity, style, finish (enamels being my favorite), insight into the eras in which they were created, and most importantly the vast array of people who collect and/or sell them. When I discovered Ebay a few years ago, I began to collect with a bit of a frenzy. It has been and continues to be an enjoyable adventure. In the end, working with Ira and Stuart on this project was just plain fun.

Chapter One
History of Lighter Development

Humanity and fire are inexorably linked. The image of a pre-historic people standing before a fire can be found in most museums that touch upon human history. Man without fire was basically an ape, living as he could off the land. Once man tamed and learned to control fire, he became a powerful entity and began to control other aspects of his surroundings. Fire enabled man to move to colder climates and survive the winters. Meat could be cooked or smoked and stored. Pottery could be fired and used to store liquids. Metals could be forged in the flames. The control of fire was a powerful force. The fire maker in a tribe was a revered person.

Jump ahead to the time of the Celts who lived approximately 2,500 years ago. The Celts were an agrarian society living mostly in northern England, Brittany, Wales, Ireland, and Scotland. In the late summer they celebrated the coming end of the harvest and the onslaught of winter. At the conclusion of the harvest ceremony, the hearth fires in each home were allowed to go out and the embers of a community bonfire were then used to relight each home's hearth fires in a ceremony to honor the new year which began with the beginning of winter.

The struggle to control fire has been a quest of humanity from its earliest days. Lighting a fire in a dry place was relatively simple, but during a long spell of rainy weather the material necessary to fuel the fire was hard to find. Creative ways were tried to carry fire or something that could start a fire. Ropes of slow burning material were formed that could be lit in a fire and would smolder for hours. This made them a convenient method of lighting a new fire elsewhere, which, in turn, could ignite a new smoldering rope. In the Iron Age it was discovered that when iron was struck against flint, it produced a shower of sparks that could be used to start a fire. Man constantly worked to invent a simple but reliable way to create fire on demand.

Many inventions came about as a result of war. When it was learned that gunpowder, when placed in a tube and ignited, would forcefully propel rock or anything moveable that could be stuffed down the tube, the gun was born. One of the earliest lighter styles was the fusee. In the seventh century it was used to fire matchlock rifles. In its simplest form, it was a flammable cord that was lit with flint and steel and allowed to smolder. Another early gun used a sparking mechanism, where a piece of flint was held against a spinning metal wheel to create sparks. This was the wheel lock rifle. The lighter is related to the early ignition systems that were used to ignite gunpowder.

An early fulminate of mercury lighter was patented by Selden and Keep in 1877. It consisted of a combustible rope attached to a metal holder that held paper caps containing fulminate of mercury. The caps exploded when struck and a burst of heat and sparks lit the rope. These caps were originally introduced in the late 1840s to mid 1850s and were used in percussion rifles and pistols.

Another early method of lighting a wick, not originally tried in weapons, used an early wet cell or chemical battery containing an acid solution with carbon electrodes. A zinc electrode was touched to the solution and the electric current was produced. The current heated a platinum wire coil, which was then exposed to a gasoline solution on a wick and produced a flame.

The average person did not fight in wars. They lived on a farm or in a town and, when they needed fire, they went to their hearth or, if their fire went out, to a neighbor's hearth. Among all the jobs that we hear of in the 1600s through the 1900s, including tailors, innkeepers, bakers, and candle makers, there is no job description for fire starter. There was little incentive to invent an easy way to create fire.

Along with the railroad, harvesting machines, cotton gins, multi-shot guns, and other important innovations of the Industrial Revolution of the 1830s to 1860s is, essentially, the modern match. Though some versions had existed since the early 1600s, none were overly practical until the "Congreve" match was invented in 1827, a strike-anywhere match invented in England in the early 1930s, and the "safety match" invented in 1844. With the invention of matches, the problem of instant fire was apparently solved.

But during World War I soldiers fighting in Europe lived and died in trenches that were wet during rainy season. They needed a reliable way to light a fire even in those adverse conditions. Matches were coated with wax to keep them waterproof, but wax covered matches did not light as easily as regular matches. Regular matches became useless in the trenches after the first rain. Slowly, lighters began to appear for use by the soldiers. To be successful, these lighters needed to be simple, to light, and to be fool-proof. Most, however, were not fool-proof, but a few were well made and easy to use. The old standard, flint and steel, was simple and with some engineering, practical for a portable lighter. The trick was making the flint soft enough and the steel hard enough to avoid rapid wearing out of the parts. By 1919, lighters using flint and steel were well established. Dunhill was one of the early makers of quality lighters. They found a ready market among the soldiers and demand for these lighters grew as other soldiers saw their benefits. After the war, the lighter's use spread worldwide.

As the technology matured in the 1920s, designers began to look at lighters and realized that they could sell them as personal men's items similar to fountain pens or watches. People began to spend money not only on lighters for everyday use, but for special designs used for "dress-up." Fashion began to dictate lighter design. The faster fashion changed, the quicker designs became outmoded and needed to be replaced. Lighters were made in precious metals and in streamlined shapes. The popularity of cigarettes had its first big boost in those same trenches of World War I where lighters got their start. By the 1930s, cigarette smoking was acceptable, encouraged, and actually considered sexy, even among women. The movies of the 1930s through the 1950s showed most stars with a cigarette in hand. Lighting a lady's cigarette with a gold or fancy lighter was a fashion statement. The cigarette lighter was here to stay – at least until smoking began to be outlawed in workplaces, eating establishments, and most other public venues.

Chapter Two
Collecting Lighters

Lighter Collecting Evolving
BY JUDITH B. SANDERS

My own collection started by accident when a friend brought me some lighters from Europe. It wasn't long before I found an interesting lighter at an estate sale and I was hooked!

It was 1983, when I first met another lighter collector; to my knowledge, there were very few of us at that time. In those early years the few lighter collectors I knew would buy any lighter they came across: Evans, novelty, Ronson, or made in Japan. Quality and condition were not very important at that time because as new collectors any lighter found was a treasure.

Since then, many more collectors have joined our ranks and collecting has become more discriminating. The words "mint" and "excellent condition" were heard more and more often and now are some of the most important words in the lighter community. With the newer collectors quickly learning more about lighters, some of the lighters we bought in the past were pushed to the back of the shelf.

Every few years, however, interests shift. Not too long ago, lighter collectors didn't even look at Scriptos. Then one collector, Guy Nishida, became interested and overnight Scriptos were hot! Evans Fruit & Eggs is another example of a peak in interest brought on by a single person's passion. I truly believe Martha Lam sparked the intense interest in Evans all by herself. The opposite can also happen. Ronson Case Lighters were extremely popular for many years, but lately interest in these seems to have dwindled somewhat.

Sometimes what differentiates popular and unpopular lighters is very subtle. "Made in Japan Lighters" are currently not very popular, but lighters "Made In *Occupied* Japan" are. Then, just to keep us on our toes, among the "Made in Japan" lighters, certain brands have suddenly become interesting – Shields and Swank for example.

Interests change with the market. At one time, no one could find a Champ Lighter from Austria and everyone wanted one. Now you can find these lighters frequently. It is the same story with the Dunhill, which was once rarely seen in collections, but now there are some glorious Dunhills in various collections. Remember when you wouldn't pick up an aluminum lighter? Currently there is a whole sub-group of aluminum lighter collectors. And, of course, you can't leave out the subject of Zippo. In the past you could buy an advertising Zippo (mint-in-the-box) for $5.00. Those days will never come again. We have yet to know how high these lovely Zippos will go.

With the ever-changing interests in the lighter collecting world, I recommend that if you have a lighter that appears to be unpopular in today's market, you just hang on and wait for the next shift or peak in interest.

So what can we look for in the future? Again, I suggest that you hang on to every lighter, whether you want it in your collection or not. As more collectors surface, some will certainly gravitate toward the less popular lighters, whether because of the lack of availability of other more exotic lighters or due to the cost of acquiring them. And when you think about it, if you are a true lighter collector – a lighter is still a lighter, whether it is worth a lot of money or not.

Some Thoughts on Collecting
BY URBAN K. CUMMINGS

Collecting things is part of the human condition and it is not just limited to humans. Think of squirrels burying food for future use. That may be a survival issue for them, but it impacts on us too. Collecting things of value can have any number of benefits. Enjoying their historic significance and intrinsic quality appeals to our senses. Great old fountain pens, watches, pipes, and cigar/cigarette lighters remind us of our past.

Naomi Hoffman, an artist friend, displaying some wisdom on the subject, believes that "curiosity plays an important role. The reason people collect things is that they learn more and more about a subject when they collect what interests them. Then every time they find another variation in the things they collect, it stimulates the brain pleasurably because the learning process produces endorphins. The human race, according to my theory, has progressed because of that sequence."

Taken another way, finding something new (or that you didn't know about), to add to your collection satisfies our curious nature and adds new knowledge. The excitement of discovery releases a feeling of pleasure. I know this for a fact because collectors often say to me they're going out hunting because "they need a fix." Sounds addictive, doesn't it?

On a more serious note, there are some guidelines that have evolved from many collectors with years' of experience in "learning the hard way." The following are worthy of mention:

Avoid the temptation to buy bargains with imperfections. With parts or leather missing, chipped enamel, or dents and dings, this is not what you want for your collection. Save your money and buy the good stuff. Years later you won't remember what you paid for it, but you will enjoy having a real quality piece. Real estate agents like to talk about location, location, location. Your emphasis should be on condition, condition, condition.

Be courageous – buck the trend. You don't have to collect the most popular (and often pricey) lighters. Case in point: Zippo Standard vs. Slims. I have 175 Slims, all with gorgeous graphics on them. They are every bit as enjoyable as the larger version and are, more often than not, substantially less expensive. Another example is flat advertisers, usually called "Flats." Previously a large bargain compared to Ronson or Dunhill, they still can be had for less money and are just as enjoyable.

Take the educational route. Buy books on your favorite subject. Learn from others' experience. Join one or more of the lighter clubs. Their conventions are more than just a learning experience. Making lighter friends and networking with fellow collectors is just plain fun. Your knowledge will grow exponentially along with your collection. With lighters becoming scarcer at antique shows and flea markets, the club conventions have become my primary source of finding lighters.

Employ a more cautious approach to your acquisitions. Ebay is an opening to the world of buyers and sellers and some great or very rare finds occur regularly. Buying from a picture and description can be hazardous at times. Getting "burned" is not uncommon. There is a lot to be said for having the piece in your hand for a careful exam and for trying it out. Live meetings and real people work the best in my opinion. But using the Internet to connect with other collectors really pays off in the long run.

A final word of caution about unmarked pieces. I have experienced this first-hand with Art Metal Works (Ronson) lighters. For some reason they were very good about putting their name on the pocket lighters they made and also on later table models. The problem occurs with their early striker lighters and their figural pieces. Some are not marked, but are known to be made by Ronson because they appear either in an advertisement or in one of the sketchbooks kept by their designer, Mr. Fredrick Kaupmann. The difficulty occurs when an unmarked striker lighter has all the earmarks of being the real thing, but there is no proof of certainty. We have no verification, so it's buyer beware. I have some of these myself. They can be a beautiful part of your collection even though they are not marked. Just enjoy them for what they are, pieces of history, and with some imagination, a look into the lives of our predecessors. Happy hunting!

A History of Lighter Collecting
BY RICHARD WEINSTEIN

This book encompasses information gathered from years of collecting and researching lighters. The many collectors who have made this book possible have freely opened up their collections to the authors to help both the novice and experienced collector alike in the search for information.

During the early days of lighter collecting, which seemed to have had its roots in the early 1960s, collecting was just a matter of buying anything that looked interesting. Price was not much of an issue since no one really valued lighters as a bona fide collectible. You could buy a lighter for a quarter and the seller was happy to be rid of it. No one knew what was common or rare, valuable or not. The joy of finding a lighter for pure collection was not generally a function of cost but one of the excitement of finding something you had never seen before. Local flea markets were the best places to look and you spent your weekends flea market hopping. If you saw a lighter that you felt was too expensive, you could be well assured that it would still be available the following weekends until the seller dropped the price to what you were willing to pay. There was little competition and you started to be known as the crazy lighter guy. Within the next few years, you started noticing that some of the lighters you didn't buy and expected to still see at the flea markets the following weekends were not showing up. Competition had started. You realized that you were not the only crazy lighter guy anymore. Someone was copying you!

As the economy grew during the 1960s and people had more disposable income, so did interest in collecting in general. Flea markets became the most popular place to hunt for all kinds of things. Little by little the flea market vendors began to notice more and more people were asking for old cigarette lighters. The introduction of collector's publications such as *The Antique Trader* gave a jump-start to collectors giving them the opportunity to find dealers throughout the United States whom they would never have had the ability to contact. Lighter collectors began running ads in the "wanted to buy" section and many "items for sale" ads began listing lighters in their columns. Soon, more and more lighters were showing up at flea markets and antique shows. Little was known about them, just that a handful of people wanted them. There were no books, no documentation to help collectors know about what they were buying, just this thirst to get something different, something they had never come across before.

There were many businesses associated with lighters, including repair services, retail stores, and manufacturers, but little thought was given to collecting the old and discarded relics of the past generations. This would all soon change. There was a need for a lighter forum and the timing was ripe for a club to take root.

Perhaps the first person to actually advertise for old lighters in a nationwide newspaper was John Cuevas of New York City, who, in 1983, placed an ad in *The Antique Trader* wanting to buy old Dunhill and Ronson lighters. At the time, lighter enthusiasts just assumed they were the only ones with this affinity. Judith Sanders of Quitman, Texas, another closet lighter collector, noticed the ad and was astounded that someone else was interested in lighter collecting as well. She contacted John to talk about lighters, as she had not known any other collectors. They began exchanging information and immediately became good friends. They frequently conversed about their new findings.

At the constant urging by John Cuevas, Judith Sanders decided to take the initiative and began the initial phase of starting a club for lighter collectors, a forum for people to share their information and help each other in the search. Finding members was a very slow process as the few interested parties were scattered around the United States (and Europe we soon learned) that it took years to capture the attention of the collecting public. The new club named "On the Lighter Side" began with three members. As more lighter collectors became aware that there were others like them, other clubs here and abroad started forming and interest in lighters grew at a rapid pace.

Lighters, once mainly purchased for their functionality, were now being collected for their beauty, design, and charisma.

In the United States, the most often asked for lighters were those made by Ronson and Dunhill, especially those made during the Deco Era – the 1920s and 1930s. The basic boom in lighter manufacturing was only fifteen or so years old by the late 1920s and Ronson and Dunhill were by far the best known and most prolific. Dunhill catered to the upper class and elite, while Ronson marketed their lighters in the moderate price range for mass appeal. No manufacturer to date had been more prolific than Ronson and they used their early Art Metal Works products to their advantage by turning many of them into beautiful, artistically decorative lighters. You may find the same figurine once used for bookends or lamps now adorned with an added ashtray and lighter. This gave the collector almost an infinite range of models to hunt for.

As collectors became more and more interested, they began to notice other lighters from different manufacturers, many of these brought by immigrants coming to America. Little was known here about the European manufacturers except Dunhill, and this was only an introduction into the vast array of lighters manufactured all over the world. There were no known books and little documentation other than some manufacturer's catalogs to help collectors understand what they were buying.

Certainly collecting was influenced by the geographic location of the collectors and manufacturers. The onset of World War II slowed the production of lighters significantly due to the rationing of metals. Some lighter manufacturers, including Ronson, turned their production to aircraft components and other military purpose, helping the businesses survive during wartime. Zippo weathered the war by supplying the armed forces with Zippo lighters made of steel instead of brass, which was needed for artillery. Many other manufacturers who could not make such changes simply vanished. Those that survived found an awaiting public thirsty for new designs. During the occupation of Japan, some companies in Japan turned to lighter manufacturing and many of their great designs are treasured by lighter collectors. Japan continues to be a major influence on lighter design

The introduction of the butane valves in the 1950s had a major influence on lighter design. The advent of this new technology gave the user a method of easily regulating the flame height as well as a much longer time between refueling. Most petrol models needed frequent refilling as the fuel would evaporate whether the lighter was used or not. The new butane models could theoretically hold fuel indefinitely if not used and could go for months without refueling under normal use. Since butane would burn with little or no smell compared to petrol, butane became the choice for pipe and cigar smokers, and continues to be the preference today.

Throughout the years, many books, including two previously released by the authors of this book, have helped collectors worldwide in their search for information. Books give collectors the knowledge needed to grow in their pursuit of lighters. The great lighter collecting books and the dedicated people who run the lighter clubs throughout the world have helped make lighter collecting the highly respectable field that it is today.

The next major boom to collecting in general must be the advent of the Internet. This monumental vehicle for the exchange of knowledge has and will forever change the way in which collectors buy, sell, and obtain information. Most notably, Ebay has given both collectors and dealers the ability to find lighters or customers from around the world as easily as from around the corner.

Lighter collectors vary greatly in what they collect. Many specialize in specific areas such as early lighter devices, precious metal lighters, strikers, enameled lighters, lighters with hidden mechanisms, Occupied Japan lighters, figural lighters depicting animals or humans, advertising lighters, or deco era lighters, or in a particular manufacturer. Whatever the area of specialization, collectors still get that wonderful feeling when they acquire a new addition to their collection.

Important Things to Consider: A Guide

Dating Lighters. As more lighter collectors have researched the lighters of the teens to the forties they have made it easier to date when lighters were made. Company records may give one date, advertisements may appear with another date, stores may have carried the lighters for years afterward. Sometimes purchasers had a lighter engraved with a date that was years after it was made. Most of the dates on the lighters in this book have a "ca." or "circa" in front of them. It means that the date of the lighter is not just limited to the date in the book. Collectors appreciate specific dates, but take each date with a period of leeway.

Condition. Condition counts. An item in mint condition may be worth several times more than one in very fine condition. Damage to a visible part may be a major problem. An item with a piece missing may be worth 25% of one with the piece present. Repair is often possible, but to repair the item, one needs parts or another of the same with that part. Ask yourself if the price is still a bargain when you have to locate and cannibalize another lighter for parts. The items in this book are priced in excellent condition although the actual piece shown may vary from good to mint in the box.

Original parts. Lighters should have original parts and finishes. In most instances, refinishing an item will not increase its value and will often decrease its value. An "honest" finish or patina, indicating that the item has seen use, is preferred.

Availability or "Will I ever find another?" Some items are always available at antique shows or specialized dealers. Ask yourself if it is just a matter of dollars to acquire an example or is this a once in a lifetime chance to find that item? With a once in a lifetime item, condition should not be the major determining factor. Many items were ephemeral or short-lived. All that is left may just be a relic or remembrance of the whole item. On those items, you may never get another chance to buy one and the person behind you may be waiting for you to put it down so that he or she can buy it.

Demand. What does it mean to you? To those growing up in the period covered by this book, some items bring back a flood of memories. To some people, seeing an item is enough. To others, possessing it is required, regardless of the price. There are collectors aplenty in the field of lighter collecting and the asking price reflects this enormous demand.

Buy the best. In comparing lighters to other collecting fields, it is expected that prices on items will continue to rise. Higher prices may be a blessing in disguise. Some people have no incentive to sell a piece for only a few dollars, but if they can get "a lot of money" they will sell it. You will pay more but you will get a lighter that you may never have another chance to own. Collections can be put together for very little money, but, as has been proven true in every field of antique collecting, the best pieces in the best condition have held or increased in value at a greater rate than the more common pieces. Buy the best that you can afford. Remember, you will rarely regret having paid too much for an item, but you will always regret the good pieces that got away.

Cost. A lighter collection need not cost a great deal of money. Some people collect newer items. Joining an organization or subscribing to a magazine can provide hundreds of leads to current items. Using your imagination, a nice collection can be put together within your budget.

Investment. We discourage people from "investing" in collectibles. Collecting should be for fun and not just profit. Profits will come if you collect the items that you enjoy and at some later time decide to sell. Buying for investment only means never playing with your things. They could get scratched or broken. Buying, selling, and trading items can help you hone a collection to those items that bring you the most pleasure. A bigger collection is not always a better collection.

Locating items. Lighters can be found at garage sales, antique stores, antique shows, flea markets, mail auctions, and through ads in the antique collector badly he wants the item. Due to increased competition, collectors on the east coast and the west coast of the United States often pay more than those in the central portions of the country.

The items in this book are valued by people who sell lighters and collectors who buy lighters. Values are given in ranges and we have tried to give values that dealers can sell at and collectors will buy at. The dealers have not set "top dollar" prices and the buyers are not paying "through the nose". It is believed that these are the fair ranges that the lighters usually trade in. The person that you deal with may not follow these "rules."

In watching the growth of lighter collecting, we have seen the interest in lighters grow from the handful of collectors in the 1970s to a full blown hobby in the 1990s that supports numerous shows, various books, a few steady dealers in lighters, loads of part time dealer/collectors and now, antique dealers who specialize in lighters. This explosion of interest may account for the increase in prices. In speaking with many collectors, it is obvious that some collectors resent the hobby's popularity and that prices have risen out of their price range. This is a natural feeling that grows from the frustration of losing a great lighter because someone else was willing to pay more for it. Often the aggravation is blamed upon a new book that lets everyone know that a certain lighter is worth a lot of money. You might hear, when asking the price of a lighter, "It's in the book at X dollars and that's what I want for it."

Valuing Lighters

In this book, we use a "supply and demand" valuation or what would a willing buyer and a willing seller agree to as a value? We have asked dealers and collectors and arrived at a value that, while not exact, is a good estimate of current value.

An examination of the components of value are helpful. A valuable item is usually rare, but a rare item is not always valuable. A rare lighter, of which only one or two are known may not have the broad appeal of a Ronson Bartender, of which there are many known. The rarity factor is not the main determinant of value. Take the example of a collector who is attempting to find a certain item for his collection. He might be willing to pay $500 for one lighter, but would he buy a second or a third at the same price? It depends upon who is buying, the availability, and how

It is often said that a price guide is out of date the moment that it is published. Do not let that affect your use of a value guide. A value guide is comparative. It allows you to compare two items to determine if they are of comparable value. It is useful in buying and trading and it can help give you a feel for the rarity of a piece.

Remember, two is a coincidence, three is a collection. Happy hunting!

Chapter Three
Lighter Manufacturers

ABDULLA. 1924 to 1940s

Abdulla made pipes, lighters, and cigarettes in France. They are extremely high quality lighters with a unique mechanism. The company was purchased by the Quercia family in the 1920s. It is believed that they were in the tobacco business prior to the 1920s. Lighters in the 1930s included models in sterling silver, gold, and hard glass enamels.

BAIER. 1930s to 1940s

Made in Germany. The postwar pieces are often marked "U.S. Zone." A high quality lighter usually seen in unpolished aluminum. Baier produced a "keg" lighter and also a couple of different jeep lighters. One is a small jeep and another is a larger jeep which holds cigarettes on its top and tows a cart behind which is an ashtray.

BEATTIE JET. 1945 to 1960s

An American-made high quality pipe lighter available with different coverings in leather, sterling silver, etc. They are fairly common in the United States but much harder to find in Europe. It had a unique system allowing one to tilt the lighter when lit. This heated a brass tube which in turn heated the liquid gasoline in the lighter and converted it into a gas. By increasing this pressure, a long "torch-like" flame was produced which was ideal for lighting one's pipe. Old Beattie Jet advertisements said this jet-like flame was ideal for small soldering jobs or for thawing out your frozen car locks.

The Beattie Jet Lighter
BY LARRY D. MARSHALL AND DEBORAH BOWMAN

Pipe lighters have taken many forms, but few have exhibited the novelty of the liquid fuel Jet lighters. The Beattie Jet lighter was invented in 1944 by Francis Leslie Phillips for Beattie Products, Inc. of New York. This device incorporated precise laws of physics that separate it from other more conventional lighters. Using a metal tube adjacent to the primary wick it produced a horizontal flame of up to two inches long.

The concept is not new and wasn't new in 1944 either, but Mr. Phillips had a very good understanding of the physics which made his device function better than any of its predecessors. According to the manufacturer's literature, this lighter could be used for lighting pipes, thawing frozen car locks, blackening gun sights, lighting fireplaces and campfires, and doing small soldering jobs.

Besides having a primary wick like most standard liquid fuel lighters, the Beattie lighter has a jet tube. Equipped with a micro orifice, the jet tube creates a torch effect. This tube is offset to the primary wick and when the lighter is tilted to allow the flame to engulf the nozzle, vaporized fuel is ignited, shooting a flame of almost two inches past the spark wheel. Within the tube is a snugly seated wick and both extend into the fuel reservoir. The internal wick stops short of the micro orifice and forms a chamber; rapidly expanding vaporized fuel is propelled through the nozzle and ignited.

The Beattie Jet lighter is an almost exact duplicate of an earlier invention (1939) by William E. Evans of the Waterbury Lock & Specialty Company in Milford, Connecticut. The differences are minor in appearance but monumental in effect. Evans' model employed exactly the same technology as the Beattie Jet but the pressure was contained in the jet tube and traveled into the fuel reservoir. This created a two-fold effect: first, excessive fuel would be forced into the primary wick and secondly, the telescoping bottom cover would allow liquid fuel to be forced up around the case and into the user's hand. This potentially dangerous situation was one of the factors that kept early model jet lighters from enjoying greater success. The Beattie Jet inventor remedied these problems by containing the pressure within the Jet tube itself. Three unique design features accomplished this: the jet tube length, the micro orifice diameter, and the jet wick density. These were calculated to provide a precise resistance and thereby disallow any pressure to be transferred to the fuel reservoir. As the fuel expanded within the jet chamber, it was expelled through the micro orifice at a determined rate. This rate is in direct proportion to the length of the flame. If an undesirable pressure was present, it could be contained within a calculated length of tubing, factoring in the density of a fuel soaked wick. If any one of these three factors does not remain constant, negative results will occur. The key to keeping the jet flame consistent and fuel leakage (due to excessive pressure) from occurring is the quality of the micro orifice. Should it become too large as a result of poor cleaning practices, the jet flame will be shortened or not light at all. If the orifice becomes partially clogged, the pressure can translate to liquid fuel leakage. Due to fouling of the micro orifice, excessive pressure can build up in all jet lighters including the Beattie. In an effort to limit this problem, the inventor designed replaceable jet nozzles and a small quantity were supplied with each lighter sold. Later versions supplied a tight diameter piano wire pick. This pick was forced into the micro orifice and would displace any carbon fouling present, without damaging the orifice. Early jet lighters had no replaceable nozzles and consequently when they stopped working, the owner would force a pin or other small object into the nozzle. As a result, the orifice would be slightly enlarged with every cleaning. As stated, the diameter of the orifice is critical to the operation of any jet lighter.

The Beattie Jet lighter has an oversized fuel reservoir to serve the combination, primary, and jet features which consume almost twice the fuel of a conventional lighter. For reference, the Jet is the same height as a standard Zippo but the thickness is greater by an eighth of an inch and the width is larger by a full quarter inch. The base metal is brass with chrome or nickel plating, creating a hefty lighter that is well built to close tolerances. The primary wick is replaceable and the manufacturer recommends the use of a long lasting Beattie glass fiber wick. Standard flints and liquid fuel work well in this model. It is not recommended to attempt changing or modifying the Jet wick or tube. Cleaning of the nozzle micro orifice should be done with extreme care.

Guy Barker of London, England invented what is believed to be the first Jet lighter in 1929. His basic design was derived from the paraffin blow lamp in which the primary flame could be extinguished once the projected flame was ignited. The physical appearance of this model is much the same as many 1920s, flip-arm types except for the conspicuous jet tube adjacent to the primary wick. Two fuel compartments with individual filling screws were provided, one for each wick. This early design minimized liquid fuel leakage.

Subsequent improvements on this design in the early and mid-thirties led to a very interesting model. Again by Francis Phillips of the Beattie Company, this unit employed a unique inner pressure chamber equipped with a bellows and pump knob, which protruded to the exterior of the case. The objective of the bellows apparatus was to allow manual pre-pressurization of the jet tube. The primary wick no longer had to be lit to pressurize. It is not known how many of these lighters were produced or exist today. The Beattie Jet lighter in its most recent form was the most successful and widely used of the Jet models. It is believed that the Beattie Jet was manufactured between 1944 and 1961. No production quantities have been found. The ultimate fatal blow to the short lived Jet lighters came in 1947. The Gas Lighter, which needed a special valve instead of a wick, was patented by a French bicycle tire valve manufacturer, Henri Pingeot in 1935 and developed during the German occupation of France by the lighter manufacturer Marcel Quercia. He launched the first gas lighter, The Gentry, in Paris, France in 1947. Jet lighters are relatively unknown, primarily due to their specialized nature, and possibly low sales or production. The Beattie Jet is the most readily available and can still be found at flea markets.

CARLTON LIGHTERS. 1918 to 1930s

The Carlton was a division of the Kum-A-Part Company, an American company which made snap buttons for clothing. The mechanisms were high quality and the enamel overlay models are very desirable. All are very collectible, and very Art Deco in appearance. The company died with the Depression.

CARTIER. 1847 to present

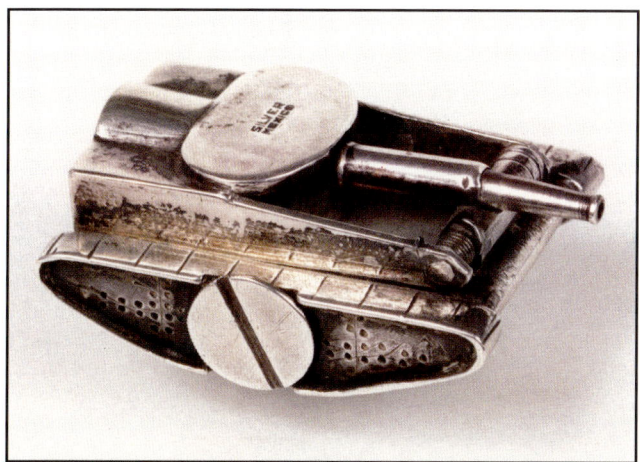

Cartier was founded in Paris, France by Louis Francois Cartier (1819-1904) who took over the shop of a jeweler named Picard. His work was of the highest caliber and he soon began to do work for King Louis Philippe and other royal and aristocrat families. His son, Alfred (1841-1925), joined him in 1872 and continued the tradition of high quality products. The firm moved to the Rue de la Paix in Paris in 1898. Alfred's son, Louis-Joseph took over control from his father around the turn of the century and began the detailed record keeping that is envied today. If you bought a Cartier product in 1900, Cartier has the records to show who made the product, when it was made, what it sold for and when and to whom it was sold. The firm was innovative; for example, the wrist watch was an invention of Louis-Joseph.

Lighter production was established after World War I. Lighters were produced before the war on an a per customer basis. Cartier produced wonderful lighters in precious metals with some containing precious gemstones, and watches You will see some of the finest and most beautiful Art Deco examples of lighters ever produced in our section of Cartier lighters.

The demand increased for their lighters during the 1920s and by the late 1920s, their products were popular with customers other than the wealthy. They used high quality production methods and their products are well made and well conceived. Cartier introduced a gas model in 1968 called the Ovale.

CLARK. 1920s to 1940s

The company was located in Attleboro, Massachusetts (home of also Marathon and Evans) and produced many beautiful and solid, high quality lighters for two decades. This included many styles and models including lighters in gold and silver, and some with built-in watches. Clark used rich lacquers and glass enamels for some of their prettiest lighters, making these very desirable to lighter collectors today. Clark also produced a few different table models including some in Sterling silver, which while scarce do turn up occasionally.

COLIBRI. 1920s to present

Originally a German firm called JBELO, which stood for Julius & Ben Lowenthal, that made pipes and lighters. The lighters were English made, however, current demand in the United States is higher than in England. In 1928, Julius developed the open spring mechanism on their "Original" model and Colibri introduced it. It is a much sought after model today, and Colibri re-introduced a replica model of their "Original" in the 1990s. In 1933, the Colibri part of the firm moved to London while JBELO stayed in Germany. Today, Colibri is still very popular selling lighters and related goods throughout Europe and North America.

DOUGLASS LIGHTERS. 1920s

Douglass made lighters for Van Cleef & Arpels and covered the market with everything from inexpensive to elegant, expensive products. Their semi-automatic swing arm action is an eye catcher, and powerfully opens for a flame. Douglass produce a variety of lighters in both Menlo Park, California as well as in the Wrigley Building in Chicago, Illinois.

The Douglass Lighter
by Ronald Eyerkuss

The Douglass lighter was developed by Leon F. Douglass about 1925 or 1926. Leon patented the lighter on October 12, 1926 and introduced it in December of 1926. Leon Douglas was somewhat of a genius in that he was credited with approximately fifty patents and was one of the founders of the Victor Talking Machine Company. It is believed that Douglas began researching lighters after he decided that matches didn't fit into the private club's "black tie" atmosphere. He wanted to have a simple and practical lighter. He tried lighters of all types and from all lands, in fact he tried for so long that he actually amassed a collection and became a cigarette lighter collector himself. None of the lighters he tried fulfilled his needs, so he made a lighter for his own personal use. This lighter was so impressive that a hard-to-please friend had to have one for himself. Douglass believed the lighter would sell to others, so he started the Douglass Lighter Company.

Leon started with one model called the "Standard". His advertising motto was "Press the trigger, there's your light". It stressed the simple and practical. The Douglass lighter began to take off when Leon found a sponsor/distributor in the Hargraft and Sons Company working out of the Wrigley Building in Chicago. Hargraft offered the lighters through private clubs, jewelers, department stores and tobacconists.

In 1928 Douglass introduced a second model, the slim "Silhouette". Douglass furthered his offerings by teaming up with the Wadsworth Company, master designers of watch cases and jewelry, to produce Douglass lighters in precious metals, such as gold and silver and then adding decorative enameling to completed the line. The lighters were offered with a slip on windshield in gold and silver finishes for use by the sportsman or outdoorsman.

A variety of finishes were available on both the Standard and Silhouette models. They included: gold, silver, gold plate, silver plate, nickel plate, chrome plate, lizard skin, ostrich, alligator, calfskin, goatskin, pigskin, water snake, tooled morocco leather, cocoa sharkskin, and pin seal.

There is a third model Douglass lighter that had what could be called a rocking chair lever action for lack of a formal name. The body of the lighter is the same as the standard Douglass but the snuffer cap and lever assembly are located in the top center of the lighter and actuate like a rocking chair motion. Pressing the thumb lever down raises the snuffer cap and lights the lighter. Little more is known about this lighter.

Douglas lighters can be identified on the bottom as being manufactured in Menlo Park, California, the Wrigley Building, Chicago, and by Wadsworth. They all should show the October 12, 1926 patent date and it is believed that there is one model that has no origin or designer name inscribed upon it. It has been rumored that the Douglass lighters are being reproduced and are showing up in Europe. It is believed that they do not contain all the inscriptions of the original.

DUNHILL. 1907 to Present

Dunhill quality is world famous and as such, their lighters are actively sought throughout the world. This is reflected in the high prices paid for some of the rare models. Dunhill was founded by Alfred Dunhill in 1907 on Duke Street, London. Before 1907, Alfred continued a horse and carriage accessories shop founded by his father. With the advent of the automobile, the shop began to make quality items for the automobile driver. One successful item (introduced in 1904) was a pipe with a built in windscreen. Alfred left the automobile accessories business and began a pipe, tobacco, and smoking accessories shop. His first Alfred Dunhill pipe, made from the best briar, was introduced in 1910. The concept of the Dunhill shop was to offer the finest products. These cost more, but Dunhill did not want to compete with other tobacco shops on price, but on quality.

Dunhill's first lighter was the "Ednite" introduced in 1914. It was a pocket tinderwick lighter. The tinderwick pocket match style lighter, followed in 1916. With the advent of World War I, soldiers became the largest business segment of sales of pipes and tobacco. Dissatisfied with the lighters on the market, Alfred wanted to sell a lighter that was easy to use and could be lit with one hand. The "Every Time" lighter was introduced in the Christmas 1923-1924 catalog. It was made for Dunhill by the English company of Wise and Greenwood. Within a year, the "Every Time" became the "Unique." Demand was so high that the Geneva firm of La Nationale was also engaged to make the lighters. One of the most sought after styles, a lighter with a watch, was not a Dunhill idea but a customer's idea. Santiago Soulas, a wealthy South American customer wanted a lighter with a watch in the side. Dunhill liked the idea and began having watch lighters made in 1926.

For clarity, we feel we should point out the various sizes of the Dunhill "Unique" models. The smallest size is the "Bijou," which measures 1.3" by 1.6". The next larger size is the "Unique A" model which is 1.5" by 1.8". Next largest size is their "Unique B," which measures 1.7" by 2.1". These are the three most common "Uniques," but there also exists a "C" model which is 2" by 2 3/8", and a "Long Bijou" which has a width of 1.9". Then there are the "Sports" models with windshields, which were sometimes incorporated into the "Bijou, A and B" models, as well as Dunhill's compendiums and "Club" desk model lighters. The "Unique Sports A" and the "Unique Sports B" Dunhills are both taller and fatter than the non-sports models.

In the 1922-1923 season, Dunhill created the Parker Pipe Company to sell pipes and smoking accessories that were not perfect enough to be called Dunhill. (Think of this as the Bentley being not quite as good as a Rolls Royce.) Parker Pipe was successful and it introduced a series of lighters sold under the name "Beacon."

Alfred Dunhill retired in 1928 and his brother, Herbert, took over control of the company aided by Alfred's daughter, Mary.

The Namiki Manufacturing Company of Tokyo made a line of high quality products with hand painted finishes. They were looking for new markets and contacted Dunhill. On July 8, 1930, Dunhill signed an exclusive agreement to sell the Dunhill-Namiki line of smoking accessories and writing instruments. On August 29, 1930, Dunhill signed an agreement with the American Safety Razor Company to market Dunhill lighters in America. Always looking to make his products better, an improved "Unique" model with a double wheel striker was introduced in 1931 along with a self winding watch lighter. A minor problem for collectors is that many customers returned their single wheel lighters to Dunhill for updating to double wheel light-

ers. This makes the single wheel lighters rarer and could be confusing when dating an early double wheel lighter that has been modified by Dunhill.

The first Dunhill "Tallboy" lighter (introduced in 1933) is unusual as it has both the Dunhill and Cartier names on it. Cartier held the patent and Dunhill held the license to produce and sell it. Dunhill's "Hunting Horn" lighter was introduced in 1934. Its "Broadboy" line first appeared in 1936 along with the "Unique Giant" lighter. 1938 saw the first "Tinder Pistol Lighter" and the "Broadboy" watch lighter came out in 1939.

Lighters were selling briskly and when the first gas lighter, called the "Gentry," was introduced in Paris in 1947, Dunhill began working on its own gas lighter. Their first model, the "Rollagas" came out in 1956.

The Dunhill "Rollalite," an export for the European market, came out in 1948. The "Sylph" first appeared in 1953, the same year as the "Aquarium." (Dunhill's records may not be accurate as the "Aquarium" is rumored to have appeared in a Christmas, 1951 catalog.) The "Sylph letter Opener" came out in 1958 and the gas "Sylphide" model followed in 1965.

The Lighter That Shouldn't Have Been
BY LARRY D. MARSHALL AND DEBORAH BOWMAN

The House of Dunhill has unquestionably been a world leader in luxury personal merchandise since 1907. Alfred Dunhill introduced his first pocket lighter, The "Unique," in 1924; advertised as the lighter that changed public opinion, it soared in popularity. For the first time, because of its revolutionary horizontal spark wheel, there was a lighter that could be operated with one hand. A host of imitations appeared on the market in subsequent years by various manufacturers, but none could compete with the quality and operability of the "Unique." To quote Alfred Dunhill, "We must try to make a better article than any other factory. This has helped us to build up our business in the past and will help us to ensure success in the future." Today, still in production, the "Unique" continues to lead in both style and craftsmanship. Its design is ageless.

The "Service" model was produced during World War II when England was under heavy siege and the Dunhill factory was partially destroyed in the blitz of 1941. If shown together with several other models (all of obvious Dunhill quality), the "Service" model is not easily recognized as a Dunhill. The government transferred most of the factory work force into war work where they used their engineering skills to make weapons for the allied forces. In addition to not being able to produce lighters, a greater problem arose as a result of the German blockade by submarine. The government commandeered all British ships and it was not possible to export any luxury goods. Only essential war weapons were carried and convoy moved these across the Atlantic.

Against this backdrop, the decision was taken to associate with companies in the United States to arrange for the manufacture of a lighter in a form to serve the needs of the allied forces at a basic cost that all could afford. Consequently, this is one of very few Dunhill pocket lighters to bear the inscription "Made in USA" just above the distinctive Dunhill trademark. The commissioning of a private United States firm to produce the "Service" model and its design criteria led to sacrifices in quality that would not have been acceptable to the House of Dunhill under any other conditions. During the war years and just following, Alfred Dunhill, Inc. of New York, set up and controlled the manufacturing aspect with an American company to manufacture these lighters. Plants in Shelton, Connecticut and Scranton, Pennsylvania produced and distributed some ten million "Service" lighters.

The "Service" model was fashioned after the classic World War I Austrian and Hungarian trench lighters. It measured 2.38" h. x 1.09" w. x 63" d. Basically, it was two cylinders formed by a single .020 thick steel stamping. The larger cylinder served as fuel reservoir and the smaller as flint magazine. The reservoir base, also a one piece stamping, unscrewed to allow refueling. Within the base cap was a piece of round leather riveted to a metal disk, which served to secure and seal the extra wick and flint compartment.

An interesting feature of the flint magazine screw was its head. Upon close inspection, it appears to be configured to resemble a daisy. It is unknown if the screw is original as no other units were available for comparison and literature is non-existent. The spark wheel is standard, much like many of today's lighters. When spun with the thumb it simultaneously lifts the wick cover, which pivots on the spark wheel axis. The fuel soaked wick is exposed to the spark and remains within the deep windproof housing formed by the fuel reservoir. Closing the wick cover smothers the flame.

The "Service" model produces an excellent shower of sparks and has proven highly reliable. Dunhill Ltd. still receives letters from elderly servicemen around the world with favorable comments. Its faults are the same as many other inexpensive liquid fuel lighters. The wick cover is not airtight and allows rapid evaporation of fuel. Additionally, the reservoir cap has no seat and does allow some leakage. The "Service" lighter may be the least expensive of the Dunhill line by far, but its collectibility is top drawer.

ELGIN. 1920s to 1950s

Elgin, the famous watch company of Illinois founded in 1835, produced many beautiful lighters from the 1920s through the 1950s. Their affiliation with "Otis," the renowned escalator and elevator company (founded in 1853), resulted in some of their fines Art Deco models. Elgin produced many watch lighters in the 1920s and 1930s with both plated and sterling silver finishes. They offered a variety of enameled combination cigarette cases with lighters, as well as some table models. Among Elgin's table models was a set with a lighter in the center and a cigarette box on either side. This was available at first with an Elgin striker lighter, and in later years with an automatic pocket lighter. They are also well-known for their many compacts produced in the 1930s and 1940s.

EVANS. 1920s to 1960s

About 1918, The Evans Novelty Company was started. They began manufacturing handbags in North Attleboro, Massachusetts. The founders were Alfred Reilly, Samuel Haslan, and Fred Burden. Evans was a metal products company that began to branch out into lady's accessories. Initially these accessories were lady's compacts. The story of the first years of this company is interesting. The Evans Novelty Company located a source that needed a supplier of compacts. Believing that they could produce compacts as well as any other company, Evans obtained the contract. At this time, however, they had no factory or machinery and had to quickly find a facility to start production. Soon, machinery was obtained, and Evans rented space in a large building that included other jewelry and accessory manufacturers. When representatives from the company that gave them the order showed up at the building unannounced, they were lead to believe that everyone in the building worked for Evans. Because of this little deceit, Evans received an advance payment that gave them enough to start production. Business was good and their product was exceptionally well made. As business continued to expand, they formed a corporation called The Evans Case Company.

The partner behind these bold moves was Alfred Reilly. He bought out his partners and put his family into the business. They continued to look for new business and by 1928 were manufacturing cigarette lighters. The lighters are often distinctive in shape and are good quality somewhat like a Ronson. This resemblance to Ronson led to another interesting story.

In 1932, Ronson (the shortened name of its founder Louis Aronson) sued Evans claiming that they had infringed upon Ronson's patented automatic lighter.

A long trial was held and the court found that Evans had indeed infringed upon the Ronson patent. Evans appealed the case to a higher court and obtained a reversal of the decision. A prosecutor heard complaints from other companies about other Appeals Court decisions. He investigated and found that the same judge had been involved in all of these cases. After an in depth investigation, it was learned that the judge had been taking bribes to decide cases in favor of those offering the bribes. Evans had paid several substantial bribes to the judge. The highest court reinstated the verdict in the lower court's trial (in this case guilty of patent infringement) and Ronson was determined to own the patents on the automatic style lighter. Ronson and Evans worked out an arrangement wherein Evans licensed the patent from Ronson and paid a royalty for every lighter made. Ronson's patent numbers can be found on Evans lighters made up to about 1952 when Evans put its own patented models into production and abandoned the Ronson patent.

The company ceased production of lighters in 1960 although lighters left in stock were sold off during the next few years.

The first model manufactured by Evans was a lift arm lighter made from 1927 to about 1929. Their first automatic lighter, the "Automatic," was introduced in about 1929. The "Automatic" lighters are usually differentiated by the "shoulder" construction. The shoulder is the top center of the lighter between the push button or lever and the snuffer. Looking from the side, the shoulder of the Automatic had an old tombstone shape. It was symmetrical, rounded on top, and then sloped down to another smaller shoulder.

Their next automatic lighter was the "Roller Bearing Action" lighter with the bulk of these being made from about 1929 through 1934. Their shoulders were symmetrical with a flattened top which then sloped down to the lighter body.

Following the "Roller Bearing Action" was the "Trig-a-lite." Their symmetrical shoulder had a flattened top, then rounded to the sides that went almost straight down without a slope. They were made from about 1932 to about 1940. There are actually a few "Trig-a-lites" made with Roller Bearing shoulders in the late 1930s.

The "Spitfire" model (an unofficial name) followed the "Trig-a-lite" which was then followed by the "Banner" model. "Spitfires" were first made in about 1940 and are characterized by the non-symmetrical sloping shoulder (what we call the typical *Evan's* look) with large screws in the side of the shoulder. The "Banner" had the same shaped shoulders but replaced the screws with pins or hid them completely. A taller, slimmer "Banner" was called the "Supreme." The "Banner" model is the lighter that is most easily recognized as an Evans.

Evans also made a "Clipper" model which did not have a push button or lever sticking up from the end opposite the snuffer. There were also a few other models and styles within a model.

There were some gas models made in the 1960s and early 1970s including a group of Evans branded Japanese Prince lighters. The lighters made in the 1930s are especially nice and collectible.

FLAMIDOR. 1907 to 1948

This was the first brand name of lighters manufactured by the Quercia family, jewelers involved in making small metal objects. They began to experiment with making lighters and in 1908 created the first "Flamidor lighter." Their last lighter was produced in 1947.

FLAMINAIRE. 1947 to 1975

French firm that made the first gas lighter. Manufacturing rights were owned by the Quercia family who owned the Flamidor company that had been making lighters in the several years just after the turn of the century. Marcel Quercia improved the model invented by Henri Pingeot in 1936. The first commercial gas lighter, the "Gentry," a table model, was introduced in June 1947. It was a very expensive lighter when first introduced. In 1948, the "Crillon," a pocket model was introduced. The company did well as the lighter was well received (it was also the only maker of the gas refill canisters). The first automatic gas lighter was the "Galet" that came out in 1959. (see also PARKER (Pen Co.)

GOLDEN WHEEL. 1920s to about 1940

Golden Wheel made a very broad range of lighters – from the cheap trinket to high quality lighters. They produced many Art Deco models as well as watch lighters, table models, and sterling silver lighters. Most were of medium quality. Collector demand is growing for these lighters.

LA NATIONALE. 1920s through present

The Swiss company, La Nationale, is an interesting company that many lighter collectors do not know about. La Nationale was started in 1905 and remained a private company until 1985 when it was sold to a Dutch corporation. They are still in business today.

Japanese Flat Lighters
BY JOHN ELSTER

Very much overlooked in the lighter world is the Japanese "Flat," as they have been nicknamed by collectors. A lighter collector who only collects high end items that have fancy mechanisms would scoff that "Flats" do not deserve any respect. After all, they are cheaply made, give-away lighters made from inferior parts. Point taken, but to the nostalgia collector who wants to collect items that truly represent the culture of the 1950s through early 1970s, these lighters are works of art that not only deserve respect, but also deserve prominent space in the best lighter collections.

What is a Japanese Flat? It is a small pocket lighter that was made from the 1950s through the 1970s. It got its name very simply: (1) it was made in Japan, and (2) it is very skinny or flat. (They measure approx. 2 1/8" long and only 3/8" wide at the widest part.) They almost always have advertising on them. Many collectors mistakenly group all Japanese advertising lighters that are not the standard Zippo size in to the "Flat" lighter category. I believe that they are wrong. Only the slim rectangular shaped ones are "Flats."

These lighters were made by a variety of manufacturers, all in Japan after the Second World War, where labor was cheap. The most prolific companies were Penguin, Rosen-Nesor, Wellington-Balboa, Rolex (no relation to the watch company), Howard, and Little Billboard. There were many others. Not all flats were manufactured the same way. There are those with advertising on only one side, those with a plastic cover over a paper advertisement, and my favorite, the metal wrapped enameled ones. These were made by taking a plain flat lighter and wrapping a piece of metal around the body. Others had colorful enameled advertising on them.

Most major companies advertised on Japanese "Flats." Some of these large companies include: Whitman's Chocolate, Levi's Jeans, Aunt Jemima, Holiday Inn, Harley Davidson, soda companies such as Coca-Cola, Pepsi, Nu-Grape, Mission Orange, and Golden Cola to name a few. There were also Flats with sports team names and logos on them such as the San Francisco Giants, the N.Y. Mets, and the Dodgers among others. There were over sixty different oil and gas companies.

Why were they made? Companies believe that these lighters were an inexpensive way to advertise their products. The lighter user would see this advertisement every time they use the lighter. They even got the nickname "pocket billboard" since they resemble the billboards that were dotting highways all over the county. Some companies gave these lighters away, outright or in return for proofs of purchase or coupons, while others sold them cheaply. The most common flats were those made by R. J. Reynolds for the Camel, Winston, and Salem cigarette brands. Research shows that there were over 24 million lighters made for these three brands alone.

Within the category of "Advertising Flats" there are many sub-categories. Some people collect only tobacco company advertising, some collect only truck advertising, soda company advertising, gas/oil company advertising, etc. You may think that if one collected only a sub-cat-

egory of flats you would not have many lighters. You would be surprised. A sizable collection can be made around any one of these sub-categories and many others.

Japanese Flats are becoming harder to find at reasonable prices – but this only makes the find that much more exciting. Flats are also a great cross-over collectible. They appeal to collectors of other items. For example a Coca-Cola collector would consider a COKE Flat a prize for their collection and a John Deere collector would feel the same about the beautiful green and yellow Flat advertising the famous tractor company. This cross-collectiblity keeps the demand for these lighters high and provides a good liquid market for sellers. A recent Ebay sale of a McDonalds Hamburger flat sold for over $400, not to a lighter collector, but to a McDonalds collector.

Valuing Flats – What makes one worth $3-$5 and another $50-$100? The condition of the lighter and the advertisement are the two things that determine the value. Since these lighters were not well made to begin with, they must be in excellent to mint condition to be worth collecting. The advertisements on the lighter are the other determining factor. A national advertisement appeals to more people so they typically sell for more than a local advertising Flat (i.e. a Flat advertising Texaco sells for more than a Flat advertising Bob's Service Station). A picture of the item that is being advertised adds to the value. (For example a flat advertising Smith Buick is nice – But one that advertises Smith Buick and shows a picture of the car is better.) Finally, the number of colors used in the advertising is important with more being better. Flats can still be found at most flea markets, antique shops, yard sales, and junk shops and at Internet auction sites.

Lighter production began in 1920. They produced lighters (the "Rollalite") for Dunhill before, during, and after World War II. They made the Ronson "Varaflame" and "Comet" lighters, (designed by La Nationale as a disposable lighter and changed by Ronson as a refillable one). La Nationale was instrumental in the success of the gas lighter. Their gas filling valve system was used by Dunhill and Ronson. In the 1950s they made lighters based upon the Ruetz system (invented in 1947 and named after Swiss inventor, Theodor Ruetz). A gas tank with a separate pressure equalization compartment was used instead of cotton wool in the body of the lighter. This increased fluid capacity and cut down on the number of parts needed.

LANCEL. 1928 to 1965

France Albert Lancel began making lighters in 1928 and continued to produce many quality lighters until 1965 The company was started by the father of a pipe maker, Albert Alphonse Lancel, in 1876. They turned out many fine pipe lighters, watch lighters, and other beautifully enameled lighters throughout the Art Deco era. In the 1950s they produced their first gas model which still resembled their well known "Automatique."

MARATHON. 1915 to 1940s

Marathon lighters are highly sought after for their Art deco look and very high quality. The earliest ones (pre-1925) are highly prized, as are their cigarette case/lighter combinations.

ORLIK LIGHTERS. 1916 to 1940s

Founded in 1899, this company began as a pipe maker. In about 1916, they moved on to making lighters, some of which were made in the U.S. and others in the U.K. Their lighters seemed to disappear in the 1940s.

PARKER (Pen Co.) 1950 to 1956

In 1950, the Parker Pen Company began looking for new products to market in order to take advantage a strong United States sales force. Because of the popularity of cigarette smoking, the company felt that a cigarette lighter was a good personal accessory to go with fountain pens. Parker bought the rights to the "Crillon" lighter made by Flaminaire, the French firm that made the first gas lighter. At the time, manufacturing rights to the Flaminaire was owned by the Quercia family who also owned the Flamidor. The Quercia family (originally jewelers) had been making lighters since about 1905.

Marcel Quercia and George Ferdinand improved the gas lighter invented by Henri Pingeot in 1936 and the first commercial gas lighter was introduced by them in June 1947. It was very expensive when first introduced. In 1948, they began to market the Flaminaire "Crillon" lighter in the United States. Because Quercia lacked a sales force in the United States, it sold the gas lighter rights to Parker in June of 1950. Parker immediately began advertising heavily for the Christmas sales season of 1950 while Quercia continued to manufacture the lighter until Parker could set up its own manufacturing facility. However, sales were not as brisk as hoped for and production was expensive. They made and marketed the lighter only until about 1956. Parker decided to leave the lighter business and dissolved the division. One reason for its poor reception was that the lighter used a single use gas canister that was only available from Parker. The Parker Flaminaire was not refillable like other later lighters on the market, thereby making the single use gas canister very expensive.

POLO LIGHTERS. 1930s to 1950s

English made, everyday working man's lighters. This lighter used a liftarm mechanism. At this time, they are relatively easy to find.

RONSON. 1909 to 1985

While most people are familiar with the Ronson Lighter Company, few know that Ronson was a leader in the decorative metal wares business well before they made lighters. The company was founded before the turn of the century by Louis V. Aronson (1869-1940) as the Art Metal Works, Newark, New Jersey (generally known as AMW). Art Metal Works began using the name Ronson (a shortened version of Aronson) in the 1920s. AMW created an incredibly diverse output of metal wares. They made bookends, automobile radiator cap ornaments, statuary, aquarium stands, plant stands, figurines, religious pieces, metal novelties, lamps, metal boxes, sparking toys, toy guns, incense burners, and, of course, cigarette lighters. Their work was of the highest quality. In the business of metal finishing, they were among the leading firms in the United States. Until the mid-1920s, the cigarette lighter business was only a tiny part of the main operation of making metal wares. As the lighter business quickly grew, it became apparent to the owners that the profits were higher on lighters than on their other metal ware lines. Management began to concentrate resources on the items that brought in the highest profits. The decorative metal wares part of their business began phasing out in the 1940s. With the advent of the Second World War, the company began producing specialized metal goods for the military. When the war ended, their decorative metal wares line was discontinued completely and the lighter, lighter fluid, and flint businesses were expanded.

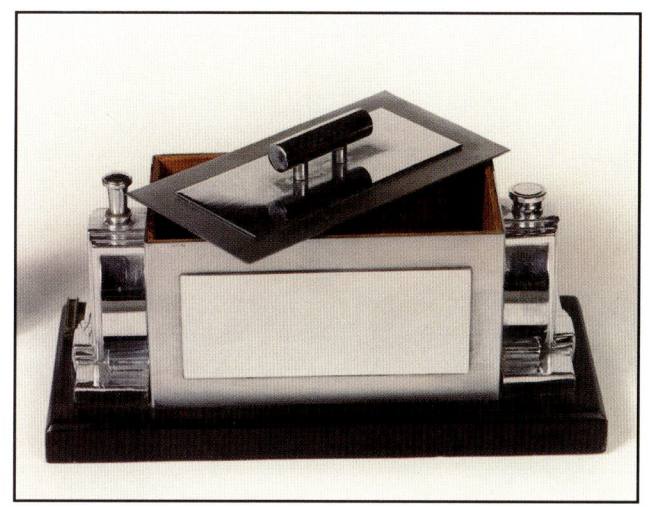

Louis V. Aronson was born of Jewish immigrant parents in New York City in December 1869. There is little information about his parents and his early years, but early on, he began to show an aptitude for things mechanical. His parents recognized his interest and enrolled him in a technical school when he was about twelve years old. He was drawn to and studied metallurgy – the science of all things metal.

The field of metallurgy is a vast domain with a long history. Think of it as starting with the copper and bronze ages, moving to the iron age, and then jumping to the industrial age, born of the American Civil War of the mid-1800s. In the early metal worker guilds in Europe, a discipline emerged that covered the science of metal composition and metal working. This included jewelry making, coin production, sword and armor manufacture, and blacksmithing. This incredible history attracted Louis Aronson's interest. He began to investigate different metal working processes. Tempering and annealing refer to the process of taking a soft metal and hardening it through heating, hammering, and cooling. Metal filling is creating a sandwich of metals that looked like the top and bottom metal but were filled with a lesser metal. An example can be found in jewelry making where the craftsman works with gold filled metal which has a copper core between layers of gold. This allows the surface to be solid gold but uses less gold to achieve the look. Casting metal is pouring molten metal into a designed mold. Metal plating is applying a thin cover of metal – gold, silver, chromium, etc. to a base metal using chemicals or electricity.

Of all these fields of metallurgy, the "emerging growth industry" during this time was electro-plating. Electro-plating was in its infancy in the 1880s and Aronson was drawn to that aspect of the business. Before this time, applying metal coatings to metal was done by a chemical process or a mechanical process that was slow, labor intensive, and produced inconsistent results. Controlling and using electricity was brand new in the 1880s. By using electricity, one could apply one metal surface to a different metal base. Louis Aronson recognized its potential and importance. Working with electro-plating, he invented the Ormolu process of gold plating in 1886. Before the Ormolu plating process, most metal makers used the gold filled process to achieve the look of gold. With gold plating, a metal worker could obtain a very similar gold look using less gold, in much less time, and using less skilled workers. The economic savings in the jewelry industry involved a great deal of money. It also allowed metal workers to create a fine jewelry finish on more items. They could now use inexpensive base metals and easily apply any number of finishes that would make the item look like they were made from a more expensive or exotic metal.

After selling his Ormolu process for a substantial sum of money, Aronson used that money to start the Art Metal Works company. The business began in New York City in the late 1880s. They later legally changed their name to Ronson Art Metal Works, Inc. in 1945 and Ronson Corporation in 1954. Their first pocket lighter in 1913 was called the "Wonderliter." It was a wand or striker type lighter (wand lighters used a metal wand with a wick. The wand was dragged across the striker and the wick would light from the shower of sparks). *Note:* Wands were often lost and replaced over time. The proper wand should fit snugly in its hole.

AMW produced its first pencil lighter 1919. It was made to look like an Eversharp style pencil but its top contained a wand. A striker was placed on the side of the pocket clip. Their first automatic cigar lighter, called the "Banjo," was introduced in 1926-1927 (automatic meaning that one push would expose the wick, turn the striking wheel and light the lighter) . After the "Banjo" came the "Standard" (1928) and then the "Princess" (1929) which was smaller. The "Princess" continued to be made for about thirty years. The early automatic lighters had exposed gears. The "Standard" and other early lighters were marked "Ronson DeLight." These used a screw that passed through the fulcrum from one side and screwed into the other side of the lighter mechanism. Later they used two screws, one entering from each side of the lighter to hold the mechanism.

In 1932 AMW sued Evans Case Company for infringing their patents and were unsuccessful until the case was reversed in 1939 when it was determined that the judge in the Evans' case was bribed to find in favor of Evans. Their patents were then licensed to Evans until the 1950s to allow them to recoup moneys lost as a result of the misuse of their patented technology.

The "Adonis" model was designed by Frederick Kaupmann and was introduced in 1947. It was made for about twenty years. Kaupmann was designing cigarette accessories fromthe late 1920s through World Wr II. Their first butane models appeared in the early 1950s. Pocket models were called "Maximus" and table models had names starting with the letter "V."

SCRIPTO. 1950s to 1970s

This Atlanta, Georgia company made Vu-lighters (see through bodies) with everything one could imagine inside. Company advertising, sports themes, fishing flies, and dice are just a few items you might find inside a Scripto lighter. As collectibles, they have become much more popular recently, although not to the extent that we have seen with Zippo lighters. They are appealing because of their great variety, and also because their price range is still reasonable. As with Zippo, salesman's samples and the advertising models are desirable, as well as political, historical, and earlier 1950s models.

S.T. DUPONT. 1940 to present

This French company was in existence before 1940, but began to make and sell lighters in 1940. The first were brass models. With the war shortages, other metals were used. Dupont introduced a gas lighter in 1953. The bottom was marked "HH" followed by four digits. In 1954, the bottom was marked "KK" followed by four digits. In 1955 and after, the code changed to "A" followed by four digits, then "B" followed by four digits, then "C "and the four digits, etc. They are known for their high quality gas models. S.T. Dupont introduced the disposable lighter, the "Cricket," to Europe in 1964 through their company Samec.

THORENS. 1920s to 1960s

Swiss made of the highest quality. They are almost in the class of Dunhills. Thorens was a music box maker

A Brief History of Scripto Manufacturing Company

BY GUY NISHIDA

The birth of Scripto Inc. in 1923 begins with a gentleman named Monie Alan Ferst. Monie's firm, M.A. Ferst Ltd., had been manufacturing lead and erasers for pencil companies since 1917. In his desire to increase sales, he became involved with the founders of Scripto (originally the Atlantic Manufacturing Co.) and its first general manager, Paul S. Hauton. Mr. Hauton was an engineer who designed the first two-way mechanical pencil capable of propelling and retracting the lead. Monie assisted Scripto in its formative years by selling their products and loaning the company money, eventually becoming its president. In fact, Monie has been credited with originating the name "Scripto." As the first U.S. manufacturer of pencil lead, obviously Monie's involvement with a competitor to his other customers could not be revealed. In a further connection, Monie was married to the sister of one of the founders of Scripto. Sometime in the late 1940s, Monie assumed majority control of Scripto in repayment of its debt to M.A. Ferst Ltd., which eventually became a subsidiary of Scripto Inc.

It was Monie's and his son Robert's connections in the pencil and lead industry that also ushered in Scripto's entry into the cigarette lighter business. The Ritepoint Company, a mechanical pencil manufacturer, had been marketing Scripto's visible fuel lighters since the late 1940s, having applied for a patent based on the Miller lighter in 1949. By 1954 Ritepoint had finally received their patent approval yet had not completely solved the problem of fuel leakage which had plagued them from the design beginnings. Rather than continue investing money in this venture, they sold the patent and rights to Robert Ferst and switched their lighter production to a Zippo-style lighter. Records indicate that sometime in 1955, the first Scripto Vu-Lighters were produced but until 1957, the appearance of the original Vu-Lighter was identical to the Ritepoint. Incidentally, in 1956 the actor Douglas Fairbanks, Jr. was on the Board of Directors of Scripto Inc.

Marketing of the early lighter models evolved in many forms, from packaging that stated Vu-Lighter Corporation as the manufacturer, then Vu-Lighter Corporation, a Subsidiary of Scripto, Inc., Vu-Lighter, Unconditionally Guaranteed by Scripto, Inc., and finally just Scripto Vu-Lighter. It is important to recognize that lighters nearly identical to Scripto except in quality were manufactured after 1977 and marketed under the name Vu Lighter, Inc.

Scripto's design and engineering staff must have been feverishly at work solving the leakage problem because by September 28, 1956, they had applied for a design patent and on February 12, 1957 had filed an application for a patent on the mechanism. Having solved the leakage problem, in large measure by reducing the number of parts from 78 to 18, the Vu-Lighter as we know it was born. In 1965 Scripto introduced the metal-clad butane lighter and in late 1969 the Vu-tane with its Lexan body was launched. Lexan was advertised as the same material used on the astronaut's helmets when they landed on the moon.

By the 1960s, Bic pens and cheap disposable lighters from abroad had contributed to and accelerated Scripto's downward spiral. To bolster profits, Scripto strayed far and wide from its writing instrument and lighter business. At one time or another, it was involved in the manufacture of carpets, ceramics, panoramic camera lenses, and copiers. Most of these ventures were ill-conceived and unprofitable diversification efforts. Through the years, Scripto sales campaigns included marketing their lighter in sets coupled with cuff links called "Lights and Links," with pens named "Write and Light," with playing cards, cigars, and other household items.

Wilkinson Sword Inc. (razor blades) purchased over half the outstanding stock in 1974. The demise of the Scripto Vu-Lighter arrived in 1977 when Scripto announced that they would cease producing lighters but they would market lighters made by licensed manufacturers overseas. By 1978, they ranked 3rd in imported lighters. In 1983, Allegheny Intl., Inc. a subsidiary of Wilkinson Sword Group Ltd., purchased all remaining shares of Scripto Inc. By 1984 Scripto-Tokai Corp of Japan took control and moved the company to California in 1986.

that began in 1883. Their most sought after lighter models are of the 1920s to 1930s era. The company moved into making record turntables, speakers, and other music items. Their "Standard Original" model was introduced in 1919 and was successful enough to convince the company to continue to manufacture lighters.

TRIANGLE. USA. 1928 to about 1940

Triangle produced mostly watch lighters both in pocket and table form. Occasionally a model without a watch will surface but this was the exception. They are very Art Deco in appearance with beveled corners and sometimes enameled watch bezels. We have seen models in chrome, silver, and gold plate and also with leather coverings.

UNKNOWN.

There are thousands of variations of lighters by unknown makers. Here are a few notes on those lighters:

French lighters made in 1911 were affixed with a copper "tax stamp" tag. The tag is dated 1911 and may have been in use for several years thereafter. During the 1920s and 1930s, French lighters were affixed with a silver "tax stamp" tag usually saying "Ministre Finance." Some were also stamped "BL" indicating a luxury tax.

Belgium lighters were required to have a metal tax stamp from 1923 into the 1930s.

The Japanese lighter industry grew dramatically in the 1950s and 1960s. Several American companies took advantage of the cheap labor in Japan and set up cigarette lighter factories. Japanese lighters made between 1945 and 1952 were stamped "Made in Occupied Japan." After the occupation, Japanese lighter manufacturers began producing thousands of models of lighters. By the end of the 1960s, the quality was among the best that could be found worldwide. Japan led the world in lighter innovation in the 1970s. With the decline in smoking, manufacturers have cut back spending the money on innovation.

ZIPPO. 1932 to present

Zippo lighter collectors are collecting social history. These lighters advertised products, goods and services, many that only existed for a short period of time. They reflect what happened during the years since the 1930s. Lighters were made to honor battles, aviation flights, space flights, award winners, and everyday events. They were used by soldiers in all the wars since the Second World War and were available in custom designs in lots as small as fifty pieces so that any group of people could memorialize their achievements. They were reliable and came with the best guarantee – If it breaks, Zippo will fix it for free. Like small clay tablets of ancient Egypt, they will tell American social history thousands of years into the future.

Prototypes vs. "Fantasy" models

There are Zippo models that exist as one of a kind. Some are prototypes, others are jeweler made while others are counterfeits. A prototype was produced by Zippo as the rough or the finished lighter that was considered for actual production. It may never have been produced or it may have been produced with changes. An offshoot to the prototype is a lighter that contained images, designed by the art department or design department, that was made to test the way the images looked on the lighter. Both of these prototypical models are rare.

Fantasy models can be described as cases designed, decorated, or created by someone other than Zippo, but they contain the Zippo made lighter mechanisms. Examples are lighters from the Vietnam war that contain obscene, anti-war, or pro-drug statements and lighters with cases made, decorated, or enameled by jewelers or Mexican silversmiths. Counterfeits are lighters that look exactly or very much like a Zippo but do not contain Zippo parts. They may be stamped with the Zippo name or a similar name such as Zippa, Flippo or Zippi or some completely unconnected name. There are collectors for each type of lighter with the prototypes being the most desirable and the high quality jeweler remade cases being the next most desirable. Most counterfeits are of a lesser quality and are sought for comparison purposes. Remember that Zippo was so successful that other companies tried to copy the image to pick up market share.

Dating Zippo Lighters

1932 - First model created by George Blaisdell. It is about one quarter inch longer than later models.

1933 to 1935 - The hinge connecting the top and bottom of the lighter case is soldered onto the outside of the case. Zippo provided models with both three and four barrel hnges. On the three barrel model, the hinge is made up of three barrels – one center, usually connected to the bottom hinge plate and two outside, connected to the top hinge plate. There were also four barrel outside hinge models produced. The bottom of the lighter case is flat and the edges are squared off. The windscreen has sixteen holes and the cap pressure bar pivot pin area is a part of the windscreen.

1936 - The hinge is still soldered onto the outside of the case and is made up of four barrels. The bottom of the lighter case is still flat and the edges are squared off.

Mid- to late 1936 to 1943 - The hinge is now soldered on the inside of the case and is made up of four barrels. The bottom of the lighter case is still flat and the edges are squared off or rounded. 1937 is the beginning of the brass drawn case with a more rounded top and bottom. Formerly, the top was flat and soldered into place.

1943-1945 - The hinge is made up of three barrels. The bottom of the lighter case is slightly rounded and the edges are rounded.

1946-1950 - The bottom of the lighter case is changed. It has a concave framed look, that is, there is an indented area where the imprint is located and the edges are rounded. In 1946, the windscreen is made with fourteen holes. In 1947, the windscreen is made with sixteen holes and the metal supporting the wheel now connects to the top of the windscreen. The words ZIPPOMFG, on the inside unit have no space between ZIPPO and MFG.

1951 forward - The hinge is made with five barrels.

1953 forward - The patent number is changed to 2517191 from 2032695.

1957 forward - There is now a code for year of manufacture on the bottom of the case as shown in the following chart.

(See chart on next page)

Dating Zippo Lighters

YEAR	REGULAR		SLIM	
	LEFT	RIGHT	LEFT	RIGHT
1932	Patent Pending			
1937	Patent 2032695 *			
1950	Patent 2517191			
1957	Full stamp with patent pending		••••	••••
1958	Full stamp, no patent pending ••••	••••	••••	•••
1959	••••	•••	•••	•••
1960	•••	•••	•••	••
1961	•••	••	••	••
1962	••	••	••	•
1963	••	•	•	•
1964	•	•	•	
1965	•			
1966	IIII	IIII	IIII	IIII
1967	IIII	III	IIII	III
1968	III	III	III	III
1969	III	II	III	II
1970	II	II	II	II
1971	II	I	II	I
1972	I	I	I	I
1973	I		I	
1974	////	////	////	////
1975	////	///	////	///
1976	///	///	///	///
1977	///	//	///	//
1978	//	//	//	//
1979	/	//	//	/

YEAR	REGULAR		SLIM	
	LEFT	RIGHT	LEFT	RIGHT
1980	/	/	/	/
1981	/		/	
1982	\\\\	\\\\	\\\\	\\\\
1983	\\\\	\\\	\\\\	\\\
1984	\\\	\\\	\\\	\\\
1985	\\\	\\	\\	\\
1986	\\	\\	\\	\\

EFFECTIVE 7-1-86 THE ABOVE SYSTEM WAS REPLACED BY YEAR/LOT CODE. YEAR IS NOTED WITH ROMAN NUMERAL/ LETTER DESIGNATES LOT MONTH (A=JAN., B=FEB. etc.)

YEAR	REGULAR		SLIM
	LEFT	RIGHT	
1986	A to L	II	SAME AS REGULAR
1987	A to L	III	SAME AS REGULAR
1988	A to L	IV	SAME AS REGULAR
1989	A to L	V	SAME AS REGULAR
1990	A to L	VI	SAME AS REGULAR
1991	A to L	VII	SAME AS REGULAR
1992	A to L	VIII	SAME AS REGULAR
1993	A to L	IX	SAME AS REGULAR
1994	A to L	X	SAME AS REGULAR
1995	A to L	XI	SAME AS REGULAR
1996	A to L	XII	SAME AS REGULAR
1997	A to L	XIII	SAME AS REGULAR
1998	A to L	XIV	SAME AS REGULAR
1999	A to L	XV	SAME AS REGULAR
2000	A to L	XVI	SAME AS REGULAR

Why Zippos?
BY ED PINEDO

Since the ingenious invention and branding by George G. Blaisdell in 1932, the Zippo lighter has become a symbol of American ingenuity. It is utilitarian, reliable, and very portable, simple yet sturdy in design, and its size and distinctive sound when snapping open and shut makes it a delight to hold and to flick.

The Zippo is not unlike a miniature canvas which has been engraved, painted, attached and covered in countless ways, reflecting the organizations, activities, people, events, and fashions over the last seventy years. In the mid 1930s, Zippos with initials and metal insignias appeared, and in 1936 the first corporate logo was produced. Since then, Zippos have been used as personalized gifts and for promotions, advertising, team building, and commemorative purposes.

People collect Zippos in many ways. Some collect Zippos with national brand advertising. Others specialize in "themes" such as in aviation, petroleum, automotive, tobacco, or beverage company advertising. Camel Zippos are very popular with some collectors. Military themes, such as ships, World War II, specific branches of military, or agencies such as NASA are popular. Zippos engraved "infield" with very personal and often colorful images and words during the Viet Nam war have a special appeal.

Zippos have been used as gifts by the White House, by movie studios to commemorate productions, by corporate executives, senior military officers, and by celebrities. They have been used to commemorate space launches, flight records, political campaigns, sports championships, birthdays, and anniversaries. Zippos have been made in gold, silver, brass, copper, steel, nickel, and aluminum. Zippos have been covered in various types of leather, paints, and finishes. Earlier Zippos were decorated by hand, and the evolution of mass production technology is reflected in the Zippo. Some collectors are intrigued by changes in the design of the case, the lighter mechanism, dates, or by design features such as the number of barrels on the hinge. Some find delight in collecting prototypes and test models.

With the growing awareness of the dangers of smoking, many organizations have sought to distance themselves from smoking related advertising. Some collectors now find delight in finding "politically incorrect" Zippos from the 1950s and 1960s that advertise hospitals, medical companies, and religious, and civic institutions, as well as image sensitive multinationals such as McDonalds, Coca Cola, and Disney.

As is with most collectibles, condition is an important determinant of value. Eye appeal, rarity, age, and type of engraving or surface treatment have an impact on the value of a Zippo. Zippos that are "cross-collectible" or sought after by collectors of other paraphernalia such as Coca-Cola collectors, can have special value. One of a kind Zippos are extremely difficult to value since they are very rarely offered for sale.

As we move into the 21st Century, smoking may eventually disappear. Yet Zippos will continue to gain popularity. The ingenious Zippo with its images and symbols often evoking memories of the past has become a piece of Americana, and making a flame with a flick and a click will always hold a special mystique.

The Town & Country Zippo
BY BOB BROCKMANN

Arguably, one of the finest Zippo lighters ever decorated was the Town & Country series. It was by far the most detailed, colorful, and certainly one of the most collectible. The idea was first conceived in the late 1940s – probably 1946 or early 1947. The defining airbrushed, baked enamel process (Town & Country) continued commercially until the late sixties. The latest designs I have seen with Town and Country artwork was on a Navy ship lighter and it was dated 1972. The reason the series was discontinued was due to new technology and the amount of work required completing a design. Surprisingly enough, the series was already being discouraged by the early 1960s.

For as much beauty and desire as the lighters have brought to the collectors, there is as much confusion surrounding what a Town & Country really is. Simply put, the Town & Country lighter is identified by whether or not the paint has been applied with an air brush. It has nothing to do with when the lighter was made or how the cavity was engraved or etched. Unfortunately, the only way to really tell what process was used is to use a jewelers loupe or magnifying glass (at least 10X magnification). With a loupe, you will need to look and see that there is "over-spray" separating the paint colors. Even if there is only two colors used in the design, you will still see how one color fades into the other. If the design is silk screened you will see the screened paint pattern and no over spray. If it has the layered screening process then you can see the different paint levels. The identification process becomes difficult and confusing because during the mid 1960s there were three different methods of applying paint; all running at the same time. They were the air-brushed, silk screened, and a layered screening method. The next article will spend more time on the different methods.

As history will have it, the Town & Country design was inspired by Jack Clark (Zippo's Art Director at the time). He was working on some illustrations for a wildlife series when they caught the eye of George Blaisdell. Mr. Blaisdell was very fascinated with nature, particularly with birds. He asked Mr. Clark to look into a new series that would have brilliant color with great detail and yet be durable enough to meet Zippo's stringent quality standards. At the same time, Mr. Clark had heard about a company in New York that had a new type of ceramic paint that was very durable. It was called Della Robba Glaze and manufactured by The Sculpture House located in New York City. Mr. Blaisdell liked the idea and asked Mr. Clark to visit the company and look into the paint.

Dale Hutton, engraver at the time, was asked to help in the design. He removed approximately .006 of an inch of metal, which created the cavity on the lighter. The first few test samples were then hand painted by brush, one color at a time. The paint also needed to be baked so Mr. Clark (until the commercial ovens arrived at the factory) took the lighters to his home and his wife baked the paint for one hour at 400 degrees in her oven. What makes the Town & Country series different from the other Zippo lighters is the paint is applied on paint. Metal was never used to separate the colors as was typical in prior years.

To get an idea of the time involved in the process, during the 1940s Zippo was manufacturing around 3,000-4,000 lighters per day. But, one person could paint just six or eight Town & Country lighters in one day and because of the drying time, it took three to four days to fully complete just one Town & Country lighter – it was a very slow and exacting process!

Once the airbrushed and baked enamel process was perfected, Mr. Blaisdell approved the method for production and the Town & Country series was offered to the public for sale in 1949 and sold for $7.50 each. Although there were hundreds, if not thousands of designs eventually produced, the first eight models originally introduced to the public for sale were the Mallard, Horse, Geese, Trout, Sloop, Setter, Pheasant, and Lily Pond. The Sailfish was eventually added sometime in the early 1950s. The Town & Country series was introduced to retailers as a "beautiful, unusual, color-bright and sales-right" lighter. They were advertised as being a "new Zip for your lighter sales" and having "sparkling bright designs that are hand engraved – hand painted with true to nature brilliance

and electro-baked ceramic paint for permanency." The designs were available on high polish full size and slim lighters. They were also available on sterling silver pocket lighters, #2, #3, and #4 Barcroft lighters but these pieces are very rare.

The Town & Country process itself changed very little during the twenty years the lighters were available. The process was to cut a design into the brass of the lighter then to spray a base coat into the cavity. After the base dried, fifteen to twenty different masks were used to create the design. The masks allowed the paint to be applied it a very specific area and prevent paint to be over sprayed onto other areas. An important distinction to recognize and one of the only significant changes made to the lighter was the method by which the cavity was created. Initially, the lighters were hand engraved with a "priest" engraving machine. Then, in 1957 the etching process was introduced and the cavity was mechanically cut. This process was performed by first etching through the chrome and nickel plating with electrolysis. Next, an acid splash technique was used to cut into the brass. At first blush, the etching process appears more complicated but there were many advantages. First, they could process fifty lighters at one time. Secondly, the etching process was considerably more accurate and improved production and profitability – by controlling the acid bath they could control depth cut into the brass. Also, the acid would "undercut" the chrome plate, which would help hold the paint in place once it was applied. This difference in engraving/etching techniques is the reason Town & Country lighters from 1949 to 1956 appear to be "deep cut" vs. the lighters painted after 1957. Again, despite this change, the paint was always air brushed one color at a time, allowed to air dry and then baked. Metal was never used to separate the colors as was typical in the past.

One additional observation relating to the Town & Country series was the new styled box that the lighter came in. Zippo put a great deal of time and money into the presentation of the lighter and the box was noticeably "richer" looking. The first box was light brown in color and had a leather type grain to the surface. It did not have a ® after the Zippo name which was on the front of the box. The lid had straight sides and did not come to a point in the middle. The inside of the lid had a gold silk-like lining. The bottom had the same simulated leather appearance on the outside as the lid and featured a red velvet lining for the lighter. The second style box was introduced around 1953 and it too had a light brown color but the company changed the appearance to a wood grain finish rather than the leather. Also, the sides of the lid came to a point in the middle and there was a ® after the Zippo name. The lid also had gold silk-like lining but the bottom of the box was burgundy on the outside (not brown) and it had the red velvet inside for the lighter to rest on.

Chapter Four
Lighters by Country

United States

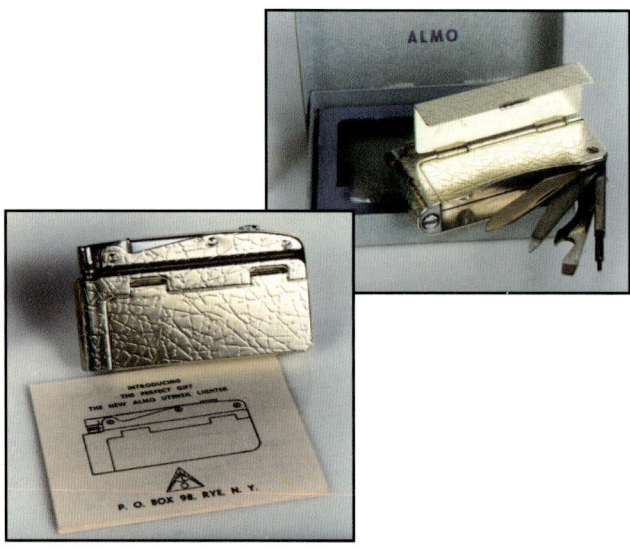

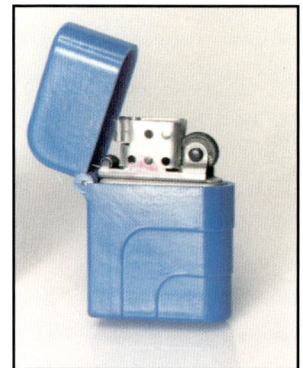

AMERICAN ROCKET PRODUCTS, ca. 1948. A lighter from a short lived company in Los Angeles, California. It has a spring loaded hinged cap and is made of a catalin type of plastic. It has that nice post war Art Deco look and has also been seen in an all metal finish marked "Rocket." 2" tall. $125-150.

ALMO, ca. 1968. A Swiss Army style of lighter with multi-purpose file, knife, opener, screwdriver, and pencil made in the USA. It is in its original box with instructions. $75-90.

ALPHA, ca. 1926. A semi-automatic, chrome plated lighter with a top loading fuel port. One of only two known to exist. 2" x 1.8". $500-750.

ARISTOCRAT, ca. 1936. A delightful semi-automatic lighter with fixed windguard. Simply twist the double wheel to the left and the lift arm is released. Powered by a spring, the arm lifts and the flint wheel spins to ignite the wick. This lighter is nickel plated brass and is marked on the bottom both "ARISTOCRAT" and "MEB" in a diamond shaped box. The fuel screw is marked "Made in U.S.A." and "FUEL." Oddly, the flint screw, which is the same size as the fuel screw (12mm) is marked "STONE." Finally the lighter is marked "PAT PEND." The importer of this lighter is "MEB." $350-450.

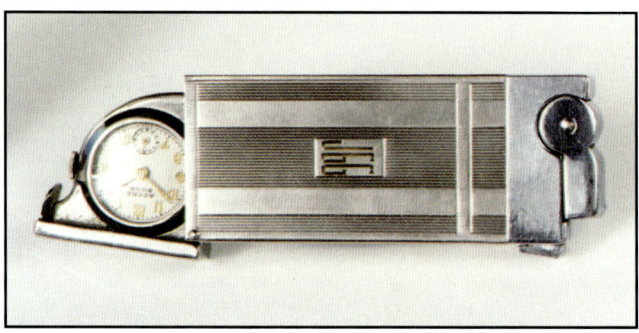

AMBASSADOR, ca. 1929. A lift arm lighter made in USA with mother-of-pearl sides. $50-75.

ASR, ca. 1948. Brooklyn, New York. A very unusual rhodium plated semi-automatic ASR lighter with a concealed Accro Bond round watch. ASR stands for American Safety Razor Corp. 2.8" x 1.2". $1,500-1,750.

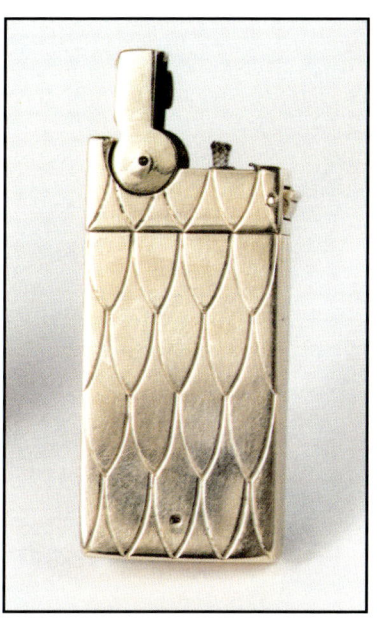

ASR, ca. 1948. A 14kt. gold ASR lighter styled by Cartier. ASR also produced a solid platinum lighter which retailed for $2,500 in the 1940s. 2.4" x 1.1". $1,000-1,250.

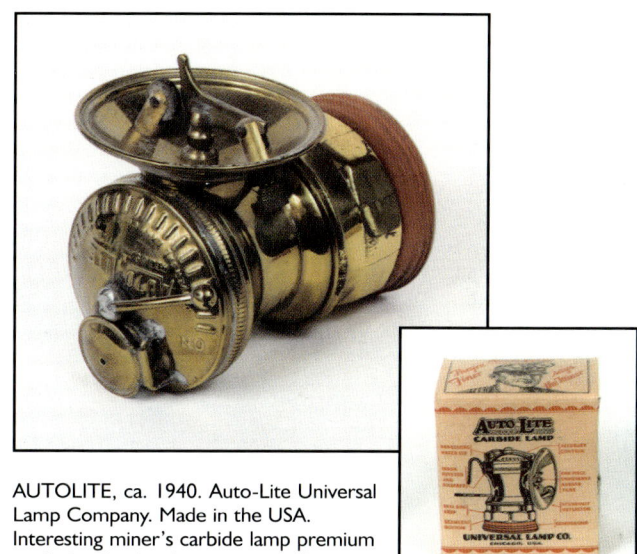

AUTOLITE, ca. 1940. Auto-Lite Universal Lamp Company. Made in the USA. Interesting miner's carbide lamp premium made for the Autolite company. The manufacturer, using existing Autolite bodies, converted them into these lighters. They were given out to customers as gifts and also displayed in various gas stations. The snuffer arm was added later and is not pictured on the box. $200-250.

ASR, ca. 1950. A chrome plated, American made, semi-automatic lighter. It has a mirror on both sides. 2.4" x 1.1". $100-125.

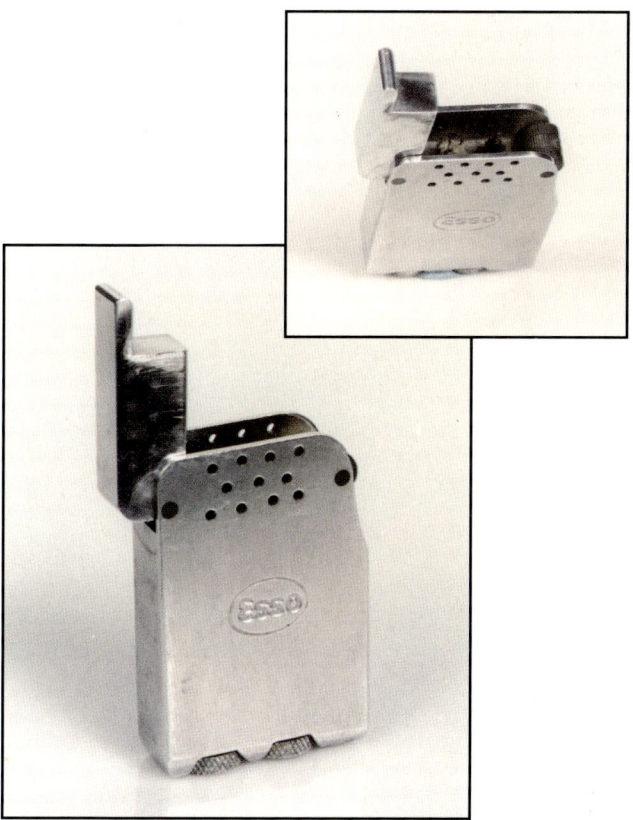

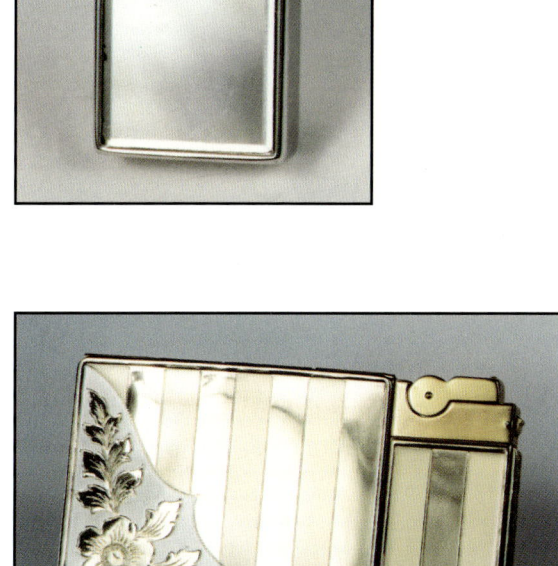

ASR, ca. 1952. American Safety Razor Co. of Brooklyn, New York. An attractive gold plated combination lighter and cigarette case. Value- $75-100.

BEACON, ca. 1949. The Beacon "Dub-L-Lite." This aluminum lighter was made in Buffalo, New York and is unusual in that it has two separate wicks designed for a larger and wider flame to allow it to light a pipe. Also nice for the Esso advertisement on the side. Also produced with a sliding bottom, called the "Hy-Grade Dub-L-Lite." 2.25" tall. $30-60.

BEATTIE JET, ca. 1942. USA made sterling silver model apparently given as a Christmas gift. $150-200.

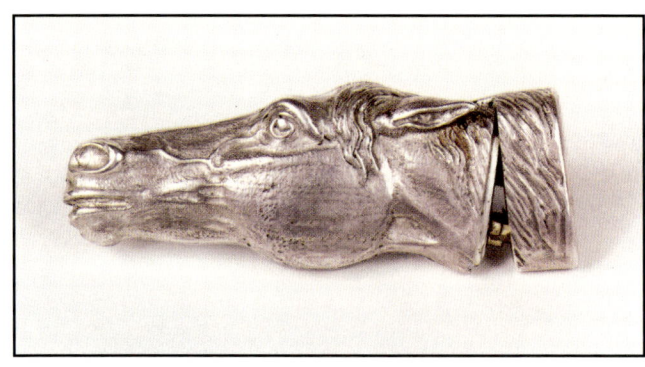

BECKWITH, ca. 1960. USA. sterling silver horse head with spring shut closure. This was a very small original production of around two dozen. Possibly patterned from a horse head match safe. Re-issued in the 1990s. 2.7" x 1". $250-500.

BEATTIE JET, ca. 1949. The Beattie Jet was an innovative lighter made for lighting pipes. When the standard wick would light and heat up the nozzle, a jet of air would blow the flame, like a small blow torch, into the bowl of the pipe. It used regular lighter fluid. $30-60.

BETTINI, ca. 1938. A chrome plated, American made, semi-automatic lighter. It has an unusual long lever in the front that activates the mechanism. Turn arm is connected to the flint wheel horizontally, to release sprung snuffer arm. 2.2" x 1.5". $1,500-2,000.

BECKWITH, ca. 1960. Beckwith Mfg. Co., La Jolla, California. The lighter on the right is a 2nd model Silver dollar made by Beckwith. The lighter on the left is unknown but surmised to be 4th or 5th model, with its cap and ring flint wheel assembly integral to the lighter. $75-150.

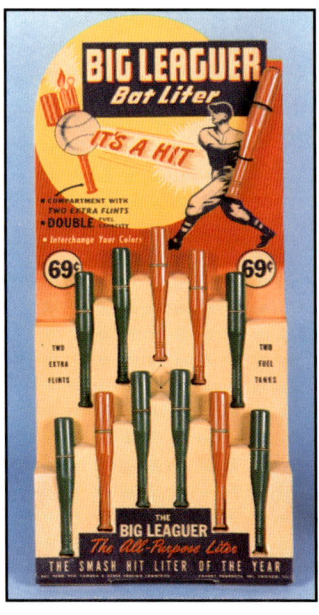

BIG LEAGUER, ca. 1940s. A lighter in the shape of a baseball bat made in Chicago. Original store display. $200-300.

BOWERS, ca. 1969. USA. This Bower's Storm Master is made of anodized aluminum. It commemorates the 1969 moon landing. $75-100.

CLARK, ca. 1924. This leather covered lighter has an uncommon windguard. 2" tall. $125-150.

CARLTON AUTOMATIC, ca. 1926. "A Kum-A-Part Product." Nickel with red leather covering. Stands 4" high. $175-225.

CLARK, ca. 1924. A nice large lift arm lighter produced in Massachusetts. 2" tall. $75-100.

CARLTON, ca. 1929. USA An automatic lighter, chrome plated with an Art Deco blue and black lacquer design. 2" x 1.4". $400-500.

CLARK, ca. 1926. A nice lizard leather "Firefly" model. 2" tall. $50-75.

CASCO, ca. 1940. The Casco electric cigarette lighter for the car probably screwed on to the ashtray. Sturdy and functional, but not attractive. $40-45.

CLARK, ca. 1926. USA. A snakeskin covered gold plated lift arm lighter. $75-100.

CLARK, ca. 1928. A leather covered table lighter that is 4" tall. $250-300.

CLARK, ca. 1928. USA. Five sided super Art Deco "Windodger" trapezoidal shaped model with incorporated windguard. Made by "Blackington" in sterling silver with enamel decoration. 2" x 1.5". $2700-3,000.

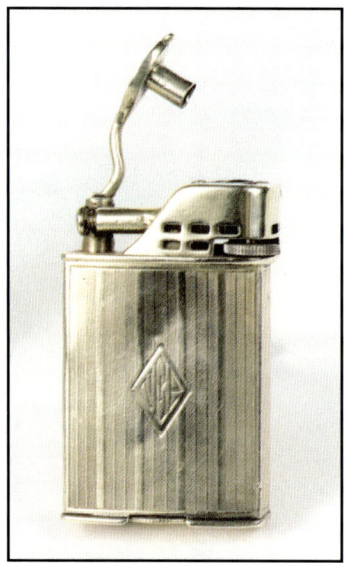

CLARK, ca. 1928. USA. A 14kt. gold model with a windguard. 2.3" x 1.5". $2,000-2,500.

 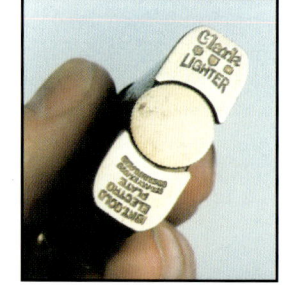

CLARK, ca. 1928. A leather covered Clark lift arm lighter with a hand painted enamel scene. The metal parts are 18kt. gold electroplated. $100-150.

CLARK, ca. 1928. USA. A six sided lighter with a three-color inlaid enamel finish. 2.1" x 1.4". $2,200-2,500.

CLARK, ca. 1928. USA. sterling silver short "Firefly" (WC initials). 1.6" x 1.4".

CLARK, ca. 1928. USA. sterling silver with enameled black stripes (WHL initials). 1.6" x 1.4". $350-450.

CLARK, ca. 1928. USA. "Windodger" model lighter. Inlaid enamel work, beveled edges, and a great tennis motif with racquet and balls. It is six sided. 2" x 1.5".

CLARK, ca. 1928. USA. Five sided "Windodger" model incorporates a windguard. Beautiful trapezoidal shape, made by "Blackington" in sterling silver. 2" x 1.5". $500-600.

CLARK, ca. 1928. USA. A "Windodger" model in sterling silver. Stylized initials "AS." 2" x 1.5". $300-500.

CLARK, ca. 1928. USA. "Windodger" model incorporates a Windguard and genuine alligator panel, nickel plated brass. 2" x 1.4". $100-200.

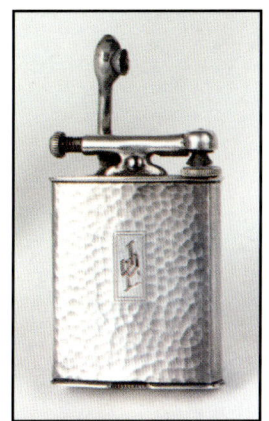

CLARK, ca. 1928. USA. Three "Firefly" models in sterling silver. Tallest 2.1" x 1.4", shortest is 1.9" x 1.4". $300-400 each.

CLARK, ca. 1928. USA. Original box and lighter with an alligator leather panel and nickel plated brass. 1.9" x 1.4". $75-150.

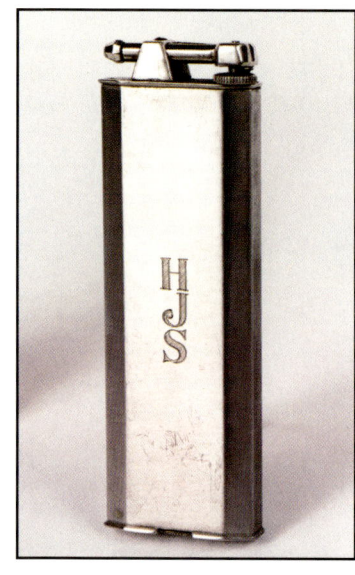
CLARK, ca. 1928. USA. A nickel-plated brass "Club" model. 4.5" x 1.4". $300-400.

CLARK, ca. 1928. USA. 14kt. cushion shaped lighter. 1.9" x 1.6". $800-1,200.

COLBY, ca. 1948. A sterling silver lighter produced in Mt. Vernon, New York by the Crosby Foundation owned by Bing Crosby. It has an interesting mechanism where a small door opens with a light when top end is pushed. Produced in two different pocket versions and also as a table model. Various enamel finishes were also available. $300-400.

CLARK, ca. 1928. USA. 14k "Firefly" model with an engine-turned design. 2.2" x 1.5". $600-800.

COLBY, ca. 1949. A USA table model with a smaller sized lighter than seen on the pocket Colby. Squeeze the end and the flap door opens to a lit wick. $125-175.

CUMMINGS, ca. 1935. Marked "J.C.N. Co. USA." Spring loaded flint wheel. Chrome plated with attractive lacquer work. $250-350.

DOUGLASS, ca. 1926. A pair of Gold-filled models. The one on the right had a windguard. The one on the left has an engine-turned design and is thinner than one on the right. This is the "silhouette" model. *Left:* $150-250; *right:* $250-350.

DOUGLASS, ca. 1926. USA. This extremely rare Douglass model, made in California, could easily be confused for a Netop. 2.4" x 1.5". $1,000-1,250.

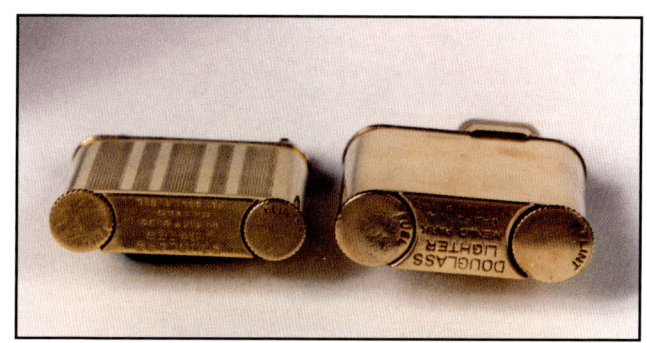

DOUGLASS, ca. 1926. An unusual and rare California made Douglass lighter with a windscreen around the wick. Height 2". $250-350.

DOUGLASS, ca. 1926. A 14kt gold-filled model marked "Wadsworth Quality". $250-300.

DOUGLASS, ca. 1926. A pair of beautiful Douglass lighters, made in California. One is solid 14kt. gold while other is gold-filled with enameling They are 2.1" x 1.6". *Left (14kt gold):* $2,500-3,000; *right (gold-filled with red and black enamels):* $1,500-2,000.

DOUGLASS, ca. 1926. A pair of very rare gold plated Douglass lighters with windguards. $250-350.

DOUGLASS, ca. 1926. A group of four table lighters with different finishes. $200-300 each.

DUO-LITER, ca. 1964. USA. A Rite-Point style see through lighter. Both the Rite-Point and the Duo-Liter were made in St. Louis, Missouri. This particular model has an adjustable wick height. Other versions do not have this height adjustment screw feature. $35-45.

ELGIN-OTIS, ca. 1929. A sterling silver lighter (shown front and back) with a watch in it. A joint venture between the Otis Elevator company and the Elgin Watch company in the 1920's. produced some fine lighters such as this. $2,000-2,500.

ELGIN OTIS, ca. 1929. Two Elgin-Otis watch lighters. On the left (with JH) is a sterling silver lighter with a "Gruen" watch. The one on the right is chrome plate with an "Elgin" watch. 2.2" x 1.7". *Left:* $1,500-2,000, *right:* $800-1,200.

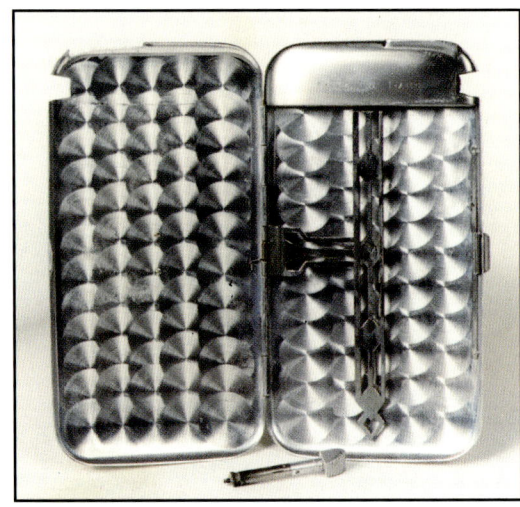

ELGIN-USA, ca. 1929. A cigarette case with a striker lighter shown in its open position. $100-175.

ELGIN OTIS, ca. 1930. Two Elgin-Otis semi-automatic lift arm lighters. The smaller is a pale yellow enamel model. The larger is a sterling silver model. Both have the same, interesting geared mechanism. The smaller lighter is 1.5" x 1.7" and the larger is 2.3" x 1.7". *Left:* $100-200; *right (sterling):* $400-500.

EVANS, ca. 1925. US made lift arm lighter with an image of a bulldog. Height 1.5". $100-150.

ELGIN-OTIS, ca. 1933. A handsome Art Deco enameled lighter with matching cigarette case. Made in USA. $200-250.

EVANS, ca. 1929. Automatic model with beautiful blue enamels. $400-500.

EVANS, ca. 1922. Enamel lift arm lighter and cigarette case with hard glass enamel in its original box. $1,000-1500.

EVANS, ca. 1930. USA. "Roller-bearing" model in a black and red enamel on chrome. $100-125.

EVANS, ca. 1930. A silver plated cigarette holder with a "Roller Bearing Action" lighter on top. The case is marked Wilcox Silver Plate Co. and was made by the International Silver Company. 5" tall. $200-250.

EVANS, ca. 1933. USA. Pretty Valerie with an Evans "Trig-a-lite" lighter with pipe rest, on a red bakelite base. Diameter of base 3.25", height 2.5". $200-250.

EVANS, ca. 1932. "Trig-a-lite" model made of a red catalin with a chrome lighter insert. $50-75.

EVANS, ca. 1933. USA. "Trig-a-lite" model with an Art Deco geometric design, made of chrome plated brass. $100-125.

EVANS, ca. 1932. Purple "Trig-a-lite." Very deco. $150-200.

EVANS, ca. 1933. A "Trig-a-lite" lighter with a very Art Deco red and black bakelite base. $150-200.

EVANS, ca. 1935. A bakelite Art Deco style lighter with a built-in clock. This is an exceptional piece in its masterful use of bakelite, color, time, and fire. Made in the USA. 6" tall. $400-500.

EVANS, ca. 1935. A brown ostrich covered Zippo type lighter. 2" tall. $125-175.

EVANS, ca. 1940. USA. "Trig-a-lite" combination lighter, cigarette case, and compact. A beautiful example of the fine quality of Evans products. $200-250.

EVANS, ca. 1935. A hinged Zippo-type lighter. 2.25" tall. $125-175.

EVANS, ca. 1940. USA. "Trig-a-lite" attractive black lacquer and chrome lighter. Notice the small ring on the snuffer to attach to a chain, if one so desired. $50-60.

EVANS, ca. 1935. A "Trig-a-lite" lighter and pen holder. The base is green and black enamel. This set was sold by Parker Pen Company around 1935. $300-400.

EVANS, ca. 1945. An unusual table lighter that was a prize for a punch board game. Two clown-like faces as shown. When the lighter portion is removed, the base is a revolving music box. $75-125.

EVANS, ca. 1936. Elephant skin covered Evans Automatic pocket lighter. $80-120.

EVANS, ca. 1945. A nice early "Banner" model cigarette case lighter. 5" tall. Shown open and closed. Evans positioned the lighter fluid filling cap outside the cigarette case in this particular model. $50-75.

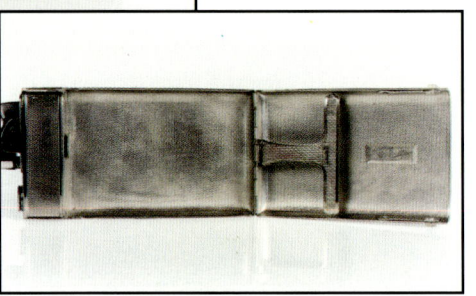

EVANS, ca. 1950. USA. "Supreme" gold plated brass lighter that is 2" tall. $40-60.

EVANS, ca. 1950. A lighter showing the USS Valley Forge. This is an unusual model with a windscreen. Height is 2". $75-100.

EVANS, ca. 1950. USA. Tall "Supreme" model with a pretty blue lacquer. Height 2.25". Missing two stones in metal frame. $60-80.

EVANS, ca. 1952. USA. Two models, "Petite" and "Banner," covered with a multi-colored mother of pearl. *Left:* 1" x 1.5" and *right:* 1.5" x 1.5". $50-75 each.

EVANS, ca. 1950. A great advertising model with the name of a radio station – WGAI Elizabeth City – on the side. $75-100.

EVANS, ca. 1952. USA. "Banner" model. Enamel lighter and matching compact in its original box. An attractive Limoges finish enameling often seen on Evans' products. $175-225.

EVANS, ca. 1952. This is an Evans "Cornucopia" model with its original box. $25-45.

EVANS, ca. 1957. A wonderful design. The painted metal cigar store Indian stands 8.25" tall and sports a "Banner" mechanism at the back of his head. $150-250.

EVANS, ca. 1952. A red enamel "Banner" model advertising Pope Precision Spindles. $125-150.

EVEREADY, ca. 1905. The Eveready flashlight company for a short time at the turn of the century produced cigar lighters. This small sterling silver engraved lighter is marked "Eveready." It was made in the USA. $75-100.

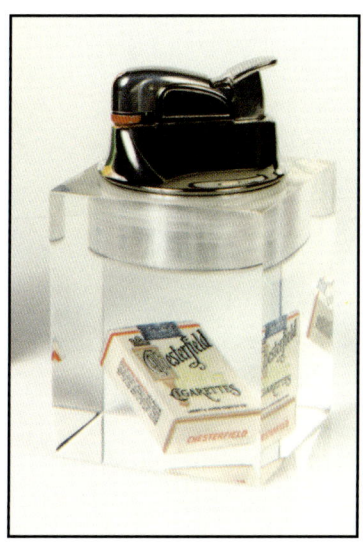

EVANS, ca. 1952. A Lucite table lighter with a miniature pack of Chesterfield cigarettes embedded inside. 3.25" tall. $100-125.

FIL-UR-LITER, ca. 1924. Dayton Pump & Mfg. Company, Ohio lighter filler made of heavy brass, advertising "Kuppenheimer Men's Clothing." Most likely found in a department store as a courtesy during the 1920s. Stands 8" high. $650-850.

FIRE-LITE, ca. 1950. A Zippo look-a-like with advertising for 7 Up, the soft drink. $25-40.

FOLMER & SCHWING, ca. 1900. A cap lighter in the form of a hunting case pocket watch. Caps are on a round disk which rotates inside the lighter. Push the watch stem slightly to release the hinged cover plate and expose the wick. Press the stem fully inward and the rotating wheel (or disc) passed under a metal scratcher which caused the fulminate of mercury to ignite. The lighter is chrome plated brass and elaborately engine-turned. Inside on the mechanism cover plate in a circular form around the edge, the lighter is marked "WATCH POCKET LAMP - FOLMER & SCHWING MF'G CO - PAT. JUNE 23, JULY 21, [18]99." Also shown is a Folmer & Schwing box of cap disks. These came in a wooden cylinder with elaborate paper labeling, containing twenty-five disks. Each disk contains thirty-six caps. Interesting information on the labels: a box cost ten cents and Folmer Mfg. was located in New York. This particular box was sold by a dealer in South Boston. The box is labeled "These Lighters are not mailable. Must be sent by Express or Freight." 2" diameter. $2,000-2,500 (with caps and box).

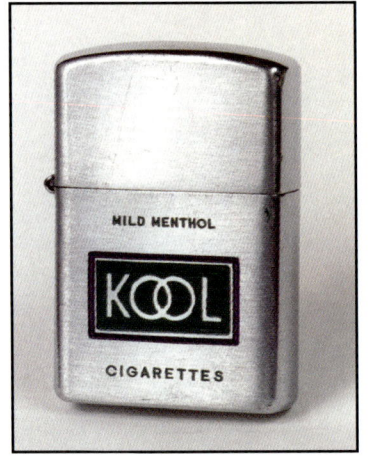

FIRE-LITE, ca. 1950. A Zippo type lighter with advertising for Kool cigarettes. $25-35.

FRANKART, ca. 1933. USA. W. C. Fields figural electric lighter. Push his red nose for a light. Frankart is known for its many beautiful Art Deco figures including statues, lamps, and ash stands. $250-450.

FISHER, ca. 1936. An Art Deco enameled lighter and cigarette case produced in Massachusetts. $400-600.

GOLDEN ARROW, ca. 1926. Lighter and case. Made in the USA. $250-350.

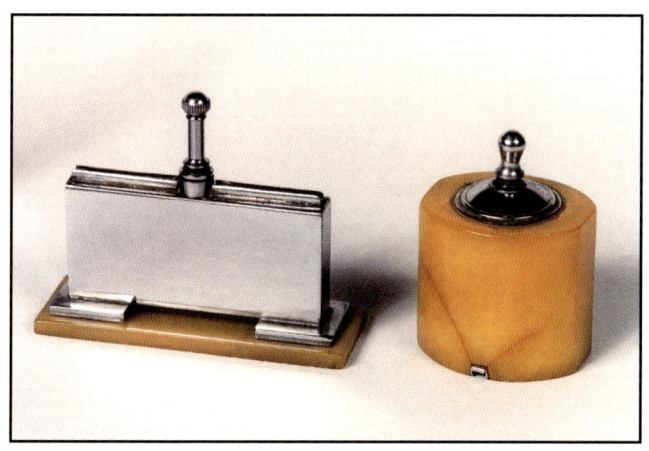

GOLDEN WHEEL, ca. 1922. Two striker lighters in catalin (plastic) and chrome. The one on the right is the "Golden Wheel" and the one on the left is unknown. $75-125 each.

GOLDEN WHEEL, ca. 1924. USA. Lift arm lighter with a diamond shaped watch. Chrome plated brass. $200-300.

GOLDEN WHEEL, ca. 1932. Lighter and cigarette case. This is a striker lighter with a beautiful enameled case. $150-200.

GOLDEN WHEEL, ca. 1928. A beautiful Art Deco set of lighter and cigarette case in their original box. They have a brown lacquer and chrome finish. $200-250.

GOLDEN WHEEL, ca. 1933. A Golden Wheel "Streamliter" in a brushed sterling silver finish. Made in the USA. $250-300.

GOLDEN WHEEL, ca. 1936. A table lighter with an automatic mechanism. Chrome plated with Art Deco base, stands 4" high. $175-275.

HELIOS, ca. 1946. An unusual bakelite lighter made in Bradford, Pennsylvania. 2" tall. $60-90.

HELIOS, ca. 1938. A Bradford, Pennsylvania-made lighter in an Art Deco bakelite model. 2.25" tall. $80-100.

J.C.N., ca. 1935. The J.C.N. lift arm lighter was made by Cummings and has enameled triangular panels. The small protruding piece on the right side of the lighter was used to turn the striking wheel. 2" tall. $300-350.

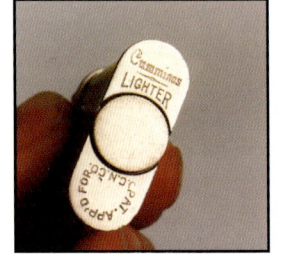

HELIOS, ca. 1940. A bakelite lighter made in Bradford, Pennsylvania. It has an attached metallique with a Texaco ad. $150-250.

KEM, ca. 1939. World's Fair Trylon & Perisphere USA with a "Kem" lighter. Chrome on wooden base. 4" across by 5" high. $150-200.

KEM, ca. 1942. USA made bowling pin lighter and cigarette case. Made of plastic in an off-white color. It pulls apart to reveal the lighter inside. Value $40-60.

KEM CO, ca. 1950. This Kem was manufactured in Detroit Michigan for the Paragon Oil Burner Co. of Brooklyn, New York. It measure 3.5" x 3.5" $60-90.

KNAPP, ca. 1924. The "Champion" model Bakelite lighter made by Knapp. 2" tall. $50-70.

KOOPMAN, ca. 1891. This is a silver plated brass semi-automatic cap lighter. Elaborate engine turning or possibly hand engraved before it was plated. A fairly common lighter, but rare in this form because of the elaborate case – all of the information usually on the outside has been moved inside. Push button marked "E.B. KOOPMAN PATENT." Inside the mechanism on the fuel tank there is the elaborate "Magic TRADE MARK" with a banner "INTRODUCTION CO. NEW YORK," then the following: "Pocket Lamp KOOPMAN'S PATS. Oct.28 89. DEC.9 90." 2.4" x 1.7". $400-600.

LEKTROLITE, ca. 1928. A nice gray speckled bakelite lighter. This was the "Plastique" model and available with a personalized silver facsimile signature. Lighter was available in a multitude of colors. 2" tall. $75-100.

LITTLE GEM, ca. 1888. An 1894 catalog page showing the Little Gem Pocket Lamp. You could buy a dozen in 1894 for $2.00. Value of actual lighter $250-300.

LITTLE GEM, ca. 1895. The Little Gem cap lighter was made in the USA. A pull down on the knob on the side of the lighter would advance a wheel to which a strip of caps was affixed. The rotating cap would pass a metal scratcher which would ignite the cap. 1.75" tall. $200-300.

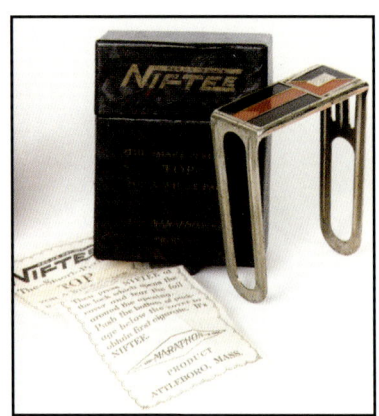

MARATHON, ca. 1936. You will not find many of these in your travels. It is a "Niftee – The Smart Cigarette Pack Top." Decorative enameled metal piece that would attach to a pack of cigarettes and cover the top of the open soft pack of cigarettes. This one was found with its box and instructions. Without them, you might not know what the item was if you saw it sitting in a dealer's case at an antique show or flea market. $50-75.

MARATHON, ca. 1934. USA. Made in Massachusetts, this lighter has a lift arm action. 2.25" tall. $60-80.

MARATHON, ca. 1936. USA. A massive windguard model with a green and black lacquer design. 2.1" x 1.5". $500-600.

MARATHON, ca. 1935. USA made red and black lacquered table lighter produced in Massachusetts. 5" high. $150-200.

MARATHON, ca. 1936. A "Super Lighter." Short and angled lift arm in nickel-plated brass. It has an interesting mechanism in that the horizontal thumb wheel turns the spark wheel. 2" x 1.5".

MARATHON, ca. 1936. An Automatic in black lacquer with tri-color gold stripes. 1.9" x 1.4". $200-250.

MATCH-KING, 1934. A Chicago "Century of Progress" Exposition striker. Also made in a pocket version depicting various exposition buildings. $150-175.

MARATHON, ca. 1936. An Automatic with a red and black lacquer design and initial shield. 1.9" x 1.4". $200-250.

MATCHLESS, ca. 1897. An early pre-flint lighter with original box and instructions. Intricate geared mechanism creates a spark with friction against a piece of "pyrite." Original box claims it is "Built like a watch, and is the ripest fruit of invention." 2.5" tall. $250-350.

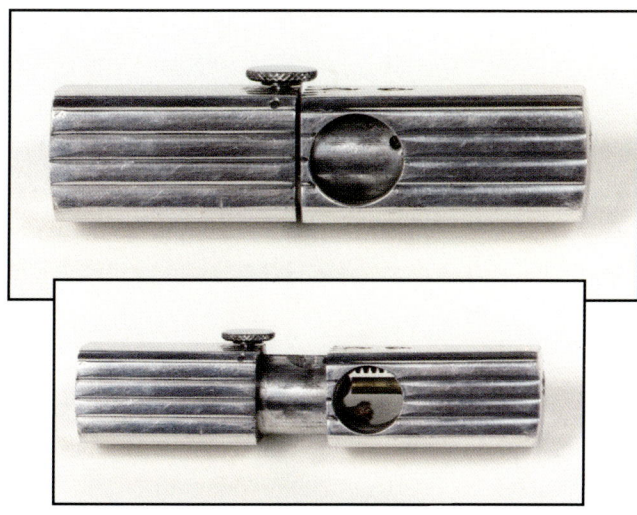

MART'S AUTOMATIC, ca. 1942. USA. Interesting spring loaded aluminum pipe lighter shown both open and closed. 4" in length. $200-300.

MAYFAIR, ca. 1933. Two Mayfair lift arms with engine-turned, sterling silver wraps. Made in USA. $150-200 each.

MEGILL'S, ca. 1880. USA. "Universal" model cap lighter in two sizes. The MEGILL's cap lighter was operated by turning the ring on the right side. This would operate the cap striking mechanism which would cause a spark to be directed against a woven tinder wick. This would then smolder and could be used to light the pipe or cigar. The larger one stores the tinder internally while the smaller one has an exposed tinder. Both are stamped REISSUE and bear four patent dates: Nov. 27, 1877, Aug. 6, 1878, Nov. 12, 1878 and May 25, 1880 (the date of reissue for these two). Both marked "MEGILL'S PATENT" and "UNIVERSAL." Chrome plated brass. *Left:* 1.6" x 2.3", $600-800; *right:* 1.4" x 2". $400-600.

NAPIER, ca. 1929. Two versions, one in silver plate on brass, the other embossed leather over brass. Both are marked on the bottom "The Napier Surefire" and "Patents Pending." The fuel screw is also marked Napier. The 1.9" x 1.8". $100-150 each.

NAPIER, ca. 1924. "Sure-Fire" model, made in the USA. Brass lift arm lighter with an intricate repoussé covering it. $75-125.

NAPIER, ca. 1936. This lighter has an interesting geared mechanism to create the spark. 2" tall. $175-225.

NAPIER, ca. 1929. A Napier trapezoidal shaped, Art Deco, lift arm lighter with enamel and eggshell on sterling silver. The bottom is marked "The Napier Lighter" and "Patents Pending." The fuel screw is marked "Napier." The 1.9" x 1.4". The red left edge of the lighter is a flaw in the photograph. $2,250-2,750.

NASSAU, ca. 1906. A sterling silver lighter with elaborate engraving. It is marked "Pat. Dec. 26, 1905." Also bears a torch-like stamp on the bottom. This model also produced in both shorter and longer size. 2.3" x 1.4". $600-800.

NASSAU, ca. 1906. USA sterling silver lighter with a plain surface and large engraved monogram. 2.3" x 1.4". $600-800.

PARK, ca. 1945. USA. A Park lighter with abalone and black lacquer decoration. $100-150.

PAIRPOINT, ca. 1932. A silver plated table model with a removable lighter insert. The Pairpoint Company was located in Massachusetts and primarily known for producing lamps during the 1930s and 1940s. $100-150.

PARK, ca. 1957. USA. An advertising model for the Amos & Andy Buick car dealership. $40-60.

PARK, ca. 1944. USA. An American Zippo-type lighter with a black crackle finish typical of those found during the war years and an ad for a jeweler. 2.25" tall. $35-50.

PARK, ca. 1970. USA. Park lighter and matching tape measure with a Good Humor Ice Cream ad. 2" tall. $35-60.

PARK, ca. 1944. USA. A black lighter with a beautiful inlaid abalone shell Oriental scene that was manufactured during the war years. 2" tall. $125-175.

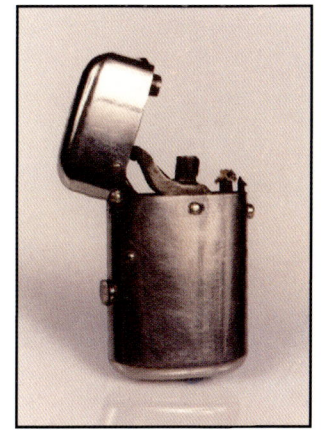

PERFECTO, ca. 1926. An uncommon American made semi-automatic pocket lighter. 2.5" tall. $125-175.

POLLACK, ca. 1928. A very rare and desirable US-made unusual lift arm lighter with a mirrored compact and rouge compartments. $400-600.

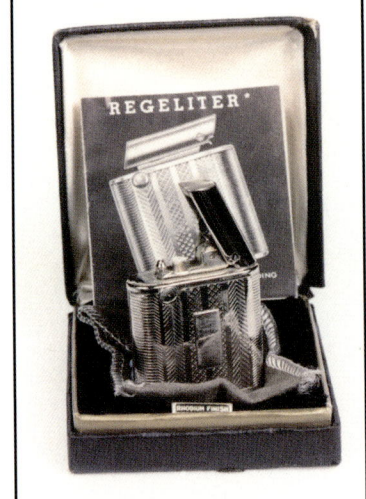

REGEL, ca. 1946. An American lighter made in Rhode Island. It is rhodium plated and has a semi-automatic action. It has a reserve fuel tank. Later referred to as "Regeliter." 2" tall. $50-75.

PRES-A-LITE, ca. 1933. A Chicago, Illinois lighter maker. A Touch-Tip lighter with a red plastic or bakelite covering. It has a replacement wand. To light requires two movements unlike Ronson's one push mechanism. It is 4" tall x 4" wide. $100-150.

REGEL, ca. 1950. The Regel was manufactured in Rhode Island. This model is rhodium plated with a push button control which both opens and ignites. This is Regel's earlier version without the bordered push button which is found in later models. It has a reserve fuel tank inside. $60-90.

PRES-A-LITE, ca. 1933. Bakelite and Chrome touch tip (a la Ronson) table lighter with clock. The clock face is marked "New Haven" and "Made is U.S.A." $600-900.

REGENS, ca. 1944. A nice group of four US made military theme pocket lighters along with their boxes, that are in unused condition. Height 2.25". $45-65 each with box.

REX-LITE, ca. 1925. USA. Lift arm lighter with a watch. Nickel plated brass with the watch crown under the snuffer arm. $200-250.

RONSON, ca. 1918. "Bison" striker lighter in an antique bronze finish, 6" in length. $1200-1600.

RONSON, ca. 1909. A pair of Art Metal Works Gobbo striker lighters with a copyright date of 1909. Gobbo was the Imp or Kewpie-type god of good luck. Both are copper plated. *Left:* 5.25" high, *right:* 6" high. $600-900 each.

RONSON, ca. 1919. A "Deer Stag" striker lighter with a bronze finish. 4" across. $1200-1600.

RONSON, ca. 1910. A wonderful "Goat" striker with a Masonic symbol. This is a very rare early Art Metal Works lighter. Length 5". $1,200-1,700.

RONSON, ca. 1919. A rare "Charlie Chaplin" striker lighter. Unmarked and similar to square based Chaplin which is a marked AMW lighter. $1500-2500.

RONSON, ca. 1920. A "Bellycan Pelican" lighter in a bronze finish. It was also made with an ashtray on a larger base. 5.75" high. $400-600.

RONSON, ca. 1919. A striker ashtray with man smoking a pipe, with Royal Bronze finish. $1000-1500.

RONSON, ca. 1921. Honus Wagner baseball player striker lighter with a pipe rest. This was also produced in polychrome color and also without a pipe rest. 5.75" high x 3.375" wide. $1,500-1800.

RONSON, ca. 1919. Very rare early striker standing 9" high. The diameter of the ashtray is 6". Bronze finish. $1,500-2,000.

RONSON, ca. 1922. An unusual Art Metal Works striker lighter with a standing dog. $600-800.

RONSON, ca. 1920. A "Bear" striker lighter with a bronze finish. 6.5 l". $1200-1600.

RONSON, ca. 1922. A nice striker made for the Mutual Oil Company of Kansas City. The barrel of oil holds the flint strip and the wand. $1,000-1,500.

RONSON, ca. 1922. "Starlight" toy sparker operates with a flint. Shown with its original box. $100-175.

RONSON, ca. 1922. The "Golf Caddy." This is a great figural Ronson striker. It was also produced with a "New Yorker" fitment. $1500-2000.

RONSON, ca. 1922. A very rare "Elk" striker lighter with an ashtray. Bronze like finish with large antlers and flint strip on his nose. $1,500-2000.

RONSON, ca. 1922. Sparking Toy (not a lighter) in the shape of a cat. It works with two flints which illuminate his eyes. Very rare. $200-300.

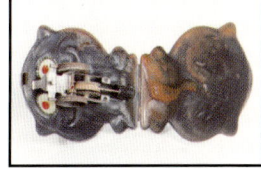

59

RONSON, ca. 1922. Art Metal Works figural striker lighter and ashtray with an early AMW plaque on its bottom. 5" across, gold finish. $500-1,000.

RONSON, ca1922. A "Cigar Store Indian" striker lighter. $900-1,200.

RONSON, ca. 1923. A woman with flowing hair. This is a very rare striker. Copper plated. 8" l. $1,000-1,500.

RONSON, ca. 1928. A 14kt gold automatic "Delight Jr." model with a small loop for a chain in front of the wick AMW stylized stamp on bottom. 1.7" x 1.5". $5,000-7,500.

RONSON, ca. 1928. A rare sterling silver model. $600-800.

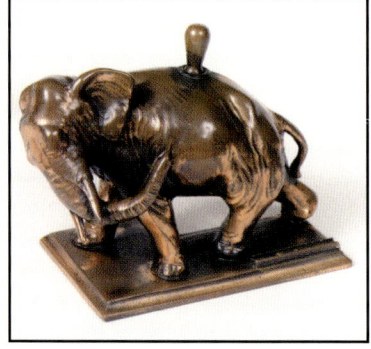

RONSON, ca. 1929. A very rare elephant striker with a copper finish and a teardrop shaped wand located in his back. Flint strip located on the base. $800-1200.

RONSON, ca. 1923. The very rare "Turbaned Arab" striker lighter. The rod is in his head and the flint is on the side of the base to his left. The bottom is marked "L.V. Aronson, copyright 1923." 4.5" tall. $1,000-1,400.

RONSON, ca. 1930. A "Perfu-Mist" with Art Deco black and white enamel stripes on chrome plated brass. It is marked on the bottom "PAT. JULY 23,1929, MADE IN U.S.A., Ronson, Perfu-Mist ART METAL WORKS NEWARK, N.J." and then (indicating this is an early one) U.S. & FOR. PATS. PDG. From a top or bottom view it looks like a Napier Surefire. Push the button down to release it. The button then rises up about .7" and becomes the pump handle. Depress the pump button and the perfume is misted and at the same time another air tube pushes the mist away from the mister. 1.8" x 1.5". $150-250.

RONSON, 1933. USA. A true Art Deco piece, Emily holds a "Cameo" model dating from the 1930s. It is chrome plated with an ivory lacquer cameo center, and a felt padded base. It stands 4" tall x 3" across. $350-450.

RONSON, ca. 1934. "Dodo Bird" striker, bronze finish. 4" w. x 3.6" h. $150-175.

RONSON, ca. 1934. A great "Dodo Bird" striker lighter with a clock. 8" l. $800-1,200.

RONSON, ca. 1934. "Sphinx" striker lighter in a black "Royal Bronze" finish. It stands 5.25" high. $1,500-2,000.

RONSON, ca. 1934. A very rare "Liberty Bell" striker lighter that was sold as a souvenir at the Pan American Exposition of California in the 1930s. $1,800-2,500.

RONSON, ca. 1934. Two geometric design striker lighters. The one on the left has a pipe rest and an inverted cone shaped wand knob. The one on the right can also be found with two pipe rests. Both stand 3" high. $600-900 each.

RONSON, ca. 1935. "Pelican" striker with a bronze finish. 5.5" h. $400-600.

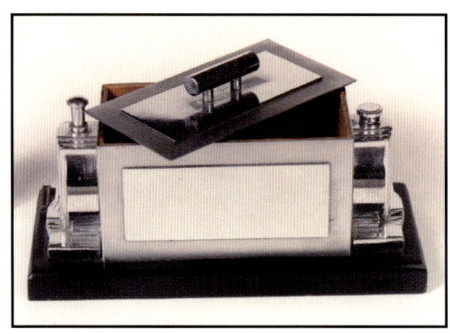

RONSON, ca. 1935. An extremely rare Ronson "Humidor Striker" lighter. A beautiful Art Deco piece measuring 10" long. This was also produced in a Touch-Tip version. Notice the strip of flint on the left side of base. $2500-3000.

RONSON, ca. 1935. A "Hound Dog With Ashtray" striker. Chrome plated on smooth black metal base. $400-600.

RONSON, ca. 1936. A Touch-Tip double cigarette holder lighter. Gold plated and faux bois finish. 3.5" tall. $300-400.

RONSON, ca. 1935. Another version of a "Hound Dog With Ashtray" striker. Smaller and more deco base than other shown. $400-600.

RONSON, ca. 1936. Two Streamlined Touch-Tip table lighters with a nice Art Deco feel. Chrome and enamel. 3.75" tall. $200-300.

RONSON, ca. 1935. Ronson's "Ballerina" striker lighter. Produced also with baby Rondelight ball lighter. $1500-2500.

RONSON, ca. 1936. USA "Touch-Tip Bartender." Undoubtedly Ronson's most famous lighter, the bartender is a true Art Deco masterpiece. Valerie proudly displays what is a miniature bar, with bartender behind preparing a cocktail. It stands nearly 7" tall x 6" across and has a depth of 2.75". The Touch-Tip lighter sits at the center of the bar, and compartments on either side hold fifteen cigarettes apiece. As the lids to these compartments are raised, the cigarettes are automatically elevated for easy access. Chromium plated with a "grained walnut-effect," and black gunmetal base. While this lighter retailed for just $25.00 some sixty-seven years ago, today's value is $2,000-2,500.

RONSON, ca. 1936. A Touch-Tip "Classic" lighter. Missing the ashtrays. Black and chrome. 8.25" long. $350-600.

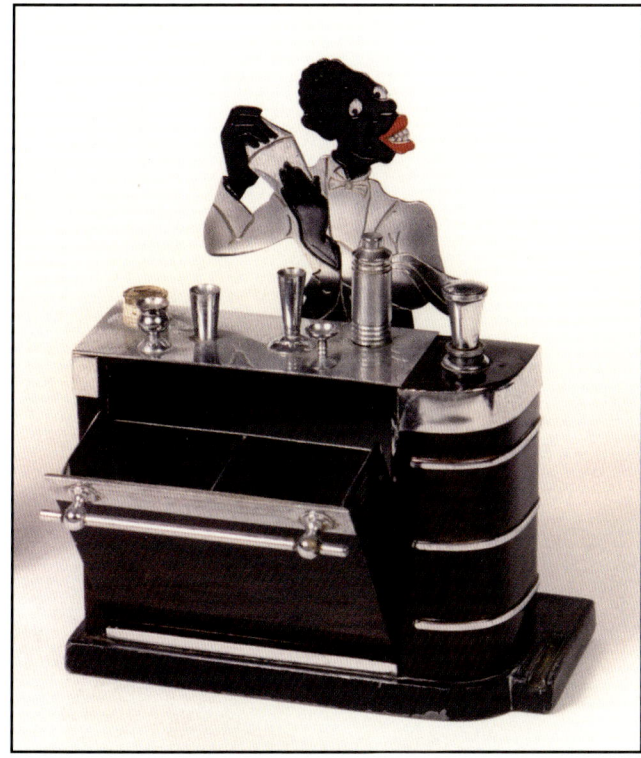

RONSON, ca. 1936. A "Touch-Tip Roll Top Cabinet" lighter. As the tray is pulled open, the chrome cover opens over the top level. Chrome and enamel. 8" wide by 6.25" deep. $500-700.

RONSON, 1936. The "Bartender" in a rare striker version. The front of the bar pulls out for cigarette storage. A strip of flint is located on the bottom right of the base. $2,500-3,000.

RONSON, ca. 1936. Very rare "Date-a-lite" lighter. The perpetual calendar has a revolving bottom so that the correct date can be set. $800-1,000.

RONSON, ca. 1936. A "Vee" with original black tassel. Main color is faux tortoise enamel, with light green enamel trim. 2.2" h. by 1.2". $350-400.

RONSON, ca. 1936. A Touch-Tip with roll-top cigarette cabinet lighter. Dureum finish with chrome. 4" tall. $500-750.

RONSON, ca. 1937. A very rare Ronson Touch-Tip humidor with pipe rests. Similar to other Touch-Tip models but this one allows for storage of tobacco inside. Gold plate and bronze finish. $1,500 to $2,000.

RONSON, ca. 1936. A Ronson Touch-Tip with pipe rests. Antique bronze finish with curved flint wheel cover and narrow tapered Touch-Tip wand. $400-500.

RONSON, ca. 1937. A combination Touch-Tip and cigarette box. Finished in chrome and enamel. 3.25" tall x 8" wide. $250-350.

RONSON, ca. 1936. A Ronson Elephant table lighter with a New Yorker fitment. Painted metal with Ivoroid tusks. 6.5" tall. $600-800.

RONSON, ca. 1937. A Touch-Tip "Smoker's Set" cigarette holder and lighter. Chrome and enamel. 6" tall. $800-1200.

RONSON, ca. 1936. A striker lighter with cigarette boxes. It is black lacquer with chrome highlights. 9.5" l. $800-1200.

RONSON, ca. 1938. A Ronson "Duette" lighter. The lighter is removable and has a ribbed collar which sat directly in a china ashtray. 4" tall. $250-350.

RONSON, ca. 1938. An American made "Banker" with an elaborate floral hard glass enamel over a silver sleeve. The back is enamel over engine turning. The sleeve is marked sterling silver with English hallmarks. Accompanied by a matching enameled case not marked Ronson. Hallmarked for Birmingham 1937-38 and bearing maker's mark HCD. The back exhibits a typical English engine turning. Lighter is 2.3" x 1.5", case is 3.3" x 4". $500-700.

RONSON, ca. 1938. A frog shaped striker with an ashtray in a bronze finish. It also has a teardrop wand. $600-1,000.

RONSON, ca. 1938. A devil head striker lighter with a copper finish. This lighter has great detail. 5" high. $1,500-2,000.

RONSON, ca. 1938. A "Touch-Tip Classic" lighter. Dureum finish with tortoise lacquer. 4.5" tall. $200-250.

RONSON, ca. 1938. A barrel shaped striker. This unusual keg shaped Ronson was also released in a "New Yorker" version. 4.5" high. $400-600.

RONSON, ca. 1938. A standing bulldog striker lighter. $400-700.

RONSON, ca. 1938. A Ronson "Touch-Tip With Pipe Holder" lighter. Bronze finish on metal. $250-450.

RONSON, ca. 1938. A "Touch-Tip Classic" with ashtray. Dureum finish with tortoise lacquer. Originally had a glass ashtray. 3.5" tall. $350-550.

RONSON, ca. 1939. A Touch-Tip and ash tray lighter. Gold plated metal. This model had a removable metal ashtray rather than the later glass ashtray. 5.5" tall. $500-700.

RONSON, ca. 1939. A "Nautical Touch-Tip" lighter with a ship's wheel. Bronze and faux bois finish. 3.25" tall. $175-275.

RONSON, ca. 1939. A "Touch-Tip Grecian Pipe Holder" lighter. Bronze finish on metal. The lighter is removable. 7.25" base. $350-600.

RONSON, ca. 1939. An "Ultracase" white enameled lighter and combination cigarette case in its original burgundy box. $125-175.

RONSON, ca. 1939. An elephant striker with a bronze finish. The striker rod which is located in his behind is not original, but a replacement from another lighter. $1000-1400.

RONSON, ca. 1939. A Ronson "Touch-Tip Cigarette Dispenser" table lighter. When the lever on the left was pushed, a cigarette rolled out. Chrome and enamel. 7.75" wide. $475-675.

RONSON, ca. 1939. A very rare "Statue of Liberty" striker lighter with a copper finish. Height is 6.75". $2,500-3,000.

RONSON, ca. 1940. A Cocker Spaniel striker lighter in a Colonial Bronze finish. The dog has a scooped out back designed for use as a pipe rest. 4.75" long. $500-800.

RONSON, ca. 1939. Extremely rare Chinese "Foo Dog" striker. Copper finish with original wand. $1000-1500.

RONSON, ca. 1941. USA. A 14kt gold automatic "Banker" model. 2.3" x 1.5". $400-500.

RONSON, ca. 1939. A Touch-Tip with cigarette holder. The cover for the cigarettes is done in plastic. Early curved flint wheel housing and tapered wand. Gold plated and enamel. 3.5" tall. $350-450.

RONSON, ca. 1941. A telephone "List Finder" striker with a bronze finish. 9" long. $2,500-3,000.

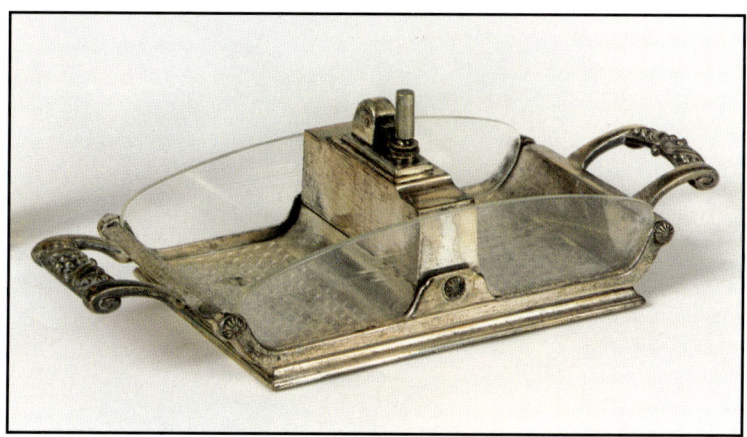
RONSON, ca. 1940. A Touch-Tip and cigarette tray set. Glass and silver plated metal. 4.5" deep by 10.5" wide. $400-600.

RONSON, ca. 1946. A wartime era "Standard" model with black enameled steel. $75-100.

RONSON, ca. 1952. This is the Ronson "Deluxe." It includes a lighter and leather wallet. This example is complete with its original box. $200-250 with box.

RONSON, ca. 1946. A wartime era "Whirlwind" model with black enameled steel. This has a concealed windscreen which can be raised to protect the flame from the wind. $75-100.

RONSON, ca. 1952. "Standard" sterling silver band over chrome. Turquoise stone in center. $150-200.

RONSON. ca. 1950. An original Christmas store display.

RONSON, ca. 1954. A Ronson "Adonis" lighter display with the lighter. $100-150.

RONSON. ca. 1950. An original store display.

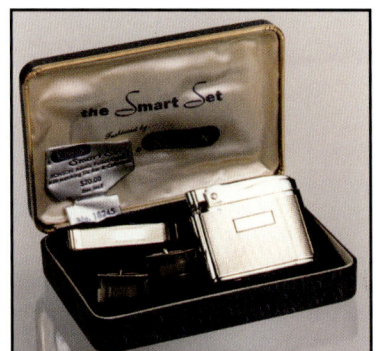
RONSON, ca. 1954. The Ronson "Adonis Smart Set." The lighter is 2" tall. This chrome set came with cuff links and a tie clip and originally cost $20.00. $75-150.

RONSON, ca. 1954. A Ronson "Masterpact" combination lighter, cigarette case, and compact store display complete with lighter. $250-300.

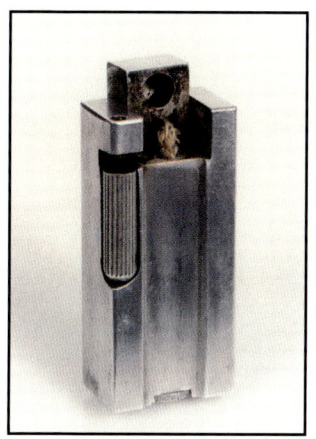
SAMPSON, ca. 1947. The Sampson lighter is a product of the industry's post-war use of aluminum. The company was located in Pittsburgh, Pennsylvania and the lighter has an unusual center-hinged snuffer. Height 2.25". $40-75.

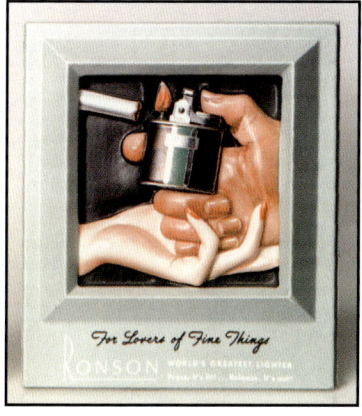
RONSON, ca. 1954. A Ronson "Standard" model tortoise and green lighter display with the lighter. $100-150.

SCRIPTO, ca. 1958. One of the earliest Vu-Lighter models advertising Randolph Fire Fighting Equipment. Red outer band. $80-100.

RONSON, ca. 1956. "Typhoon" lighters with advertising. 2" tall. $25-40 each.

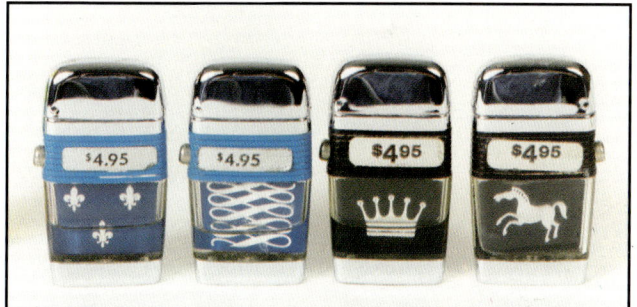
SCRIPTO, ca. 1958. Four Scripto Vu-lighters with black or blue bands and various designs inside. $30-60 each.

RONSON. A great group of miscellaneous Ronson lighters made from the 1930s to the 1950s.

SCRIPTO, ca. 1958. Four Scripto Vu-lighters with black or blue bands and outdoor sports theme designs inside. $25-45 each

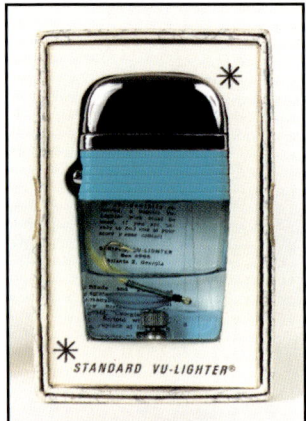

SCRIPTO, ca. 1962. A Scripto Vu-Lighter with a turquoise blue band and a fishhook inside. Note that the fluid reservoir is also a blue color on this model. $40-80.

SCRIPTO, ca. 1960. Three Vu-Liter lighters – the Pepsi Cola advertising lighter is the most valuable. The lavender "I'm a Pussycat" is the most unusual color. 2.5" tall. With box, $50 to $125 each.

SCRIPTO, ca. 1963. Vu-Lighter with an airplane inside and with blue outer band. $40-60.

SCRIPTO, ca. 1960. Three different Scripto Vu-Lighters with sports themes. A duck for the hunter, a bowler, and a golfer grace the inside of these lighters. $30-45 each

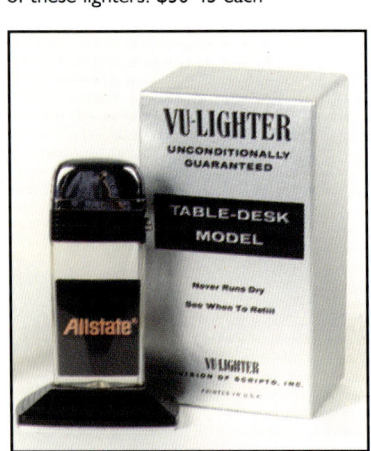

SCRIPTO, ca. 1962. A Table Vu-Lighter with its original box. The lighter contains an Allstate Insurance Co. ad insert. $80-120.

SCRIPTO, ca. 1963. Vu-Lighter table model advertising one of America's earliest hamburger chains, Carrol's. The sign reads "Carrol's 15 cent Hamburgers." $150-200.

SCRIPTO, ca. 1964. A mint Scripto Vu-Lighter with a sparkle pattern insert. $40-75.

SCRIPTO, ca. 1965. Compact Vu-Lighter with airplane. Red outer band. $30-40.

SCRIPTO, ca. 1964. Vu-Lighter showing Donald Duck with "WDP" (Walt Disney Pictures) beside him. $300-400.

SCRIPTO, ca. 1967. Vu-Lighter Sports Series with "Curler" inside. Black band. $50-80.

SCRIPTO, ca. 1965. Vu-Lighter with a green band and a tennis player inside. One of Scripto's Sport Series models. $30-40.

SCRIPTO, ca. 1967. Vu-Lighter with great advertisements for "Snowy Bleach, Glass Wax, and Mr. Bubbles. Black outer band. $50-100.

SCRIPTO, ca. 1965. Vu-Lighter with a Masonic seal and its hard-to-find box. Blue outer band. $40-60.

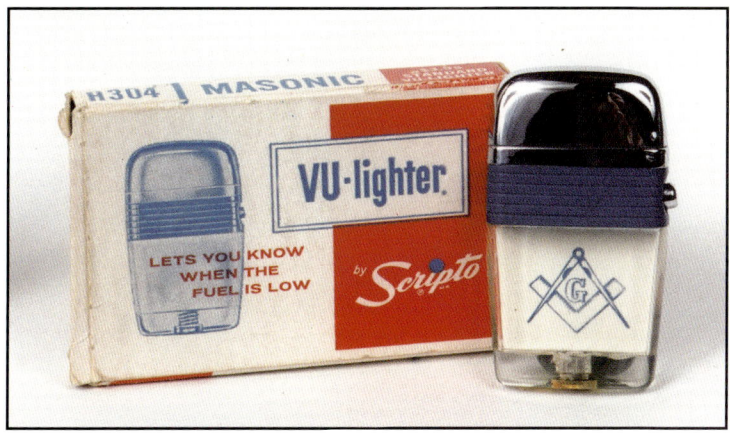

SCRIPTO, ca. 1967. Vu-Lighter with a white outer band and with "Member – Oldsmobile Somebody Loves Me Club" inside. $50-75.

SCRIPTO, ca. 1970. Vu-Lighter salesman's lighter reading "Vu-Lighter For Effective – Lasting Sales Promotion and Business Gifts." $75-125.

SCRIPTO, ca. 1970. Vu-Lighter compact "Goldenglo" model with a gold finish. It shows a television ad for Channel 17. $30-50.

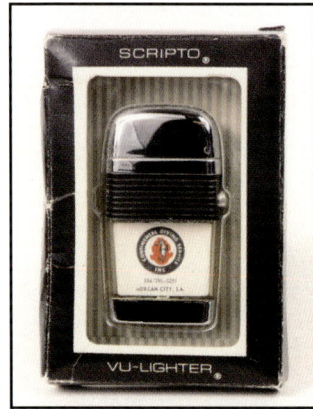

SCRIPTO, ca. 1970. Vu-Lighter made in Atlanta, Georgia showing an ad for Continental Diving Services, Inc., Morgan City, Louisiana. Black outer band. $50-75.

SCRIPTO, ca. 1970. Vu-Lighter with a black outer band and an advertisement for "Hollywood Light Bread" inside. $50-75.

SCRIPTO, ca. 1974. Vu-Lighter in a later Scripto butane box. Red band with "World's Greatest Golfer" inside. $40-60.

SCRIPTO, ca. 1970. A pocket Vu-Lighter with its original box. The lighter contains a psychedelic op-art design insert. $40 to $80.

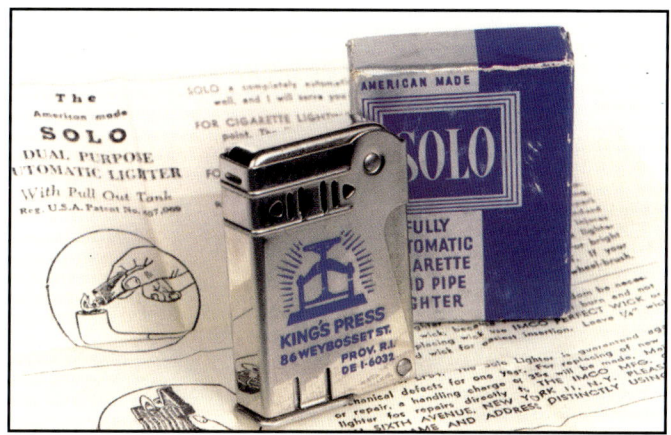

SOLO, ca. 1952. An American made lighter with an advertisement for King's Press of Rhode Island. 2.25" h. $35-55.

SCRIPTO, ca. 1975. Vu-Lighter compact "Goldenglo" model with a gold finish and black fishing fly inside. $50-75.

TOP HAT, ca. 1937. A US made lighter with its original box. Its safety feature was that you had to push the buttons on the front and back to make it operate. Chrome plated with a nice design to grip with slippery fingers. $125-150.

STRIKALITE, ca. 1933. USA. Three similar Catalin table lighters. The red one is branded "Strikalite NYC." The large one is an advertiser: "Kalt Lumber Co., Ring Regent 3100" is engraved on the bottom. The small one is simply marked "Japan" on the top. Both the red one and the small one are semi-automatics. The large one is entirely manual. $100-200 each.

TRIANGLE, ca. 1927. A US made, lift arm, watch lighter. Chrome plated with blue enamel bezel. Marked "Patent Pending." Available in many different finishes and styles of dials. Many are marked with a September, 1928 patent date, indicating this is one of their earlier models. 1.8" x 1.6". $200-300.

STRIKALITE, ca. 1949. USA. Emily enjoys this "Statue Of Liberty" lighter produced by Strike-a-lite Ltd. Inc. more than fifty years ago. A standard wheel and wick insert appropriately sits where Lady Liberty's torch normally would. 9" h. $100-200.

TRIANGLE, ca. 1928. A US made, lift arm watch lighter. Gold plated brass with beveled edges. Marked "Pat. Sept. 1928." 1.8" x 1.6". $200-250.

Unknown, ca. 1885. A dog cigar lighter. This was a pre-flint "permanent match" lighter. It was lit with a match for a continuous flame. Fluid was added by opening its hinged hat. 3" h. $200-300.

Unknown, ca. 1910. A very interesting "Bison" striker in a Copper Plate. 4" h. x 4" l. $275-475.

Unknown, ca. 1907. 14kt gold, US made catalytic lighter. It used a special denatured alcohol for fuel. Reaction to the air caused the tiny platinum balls at the end of the metal rods to produce a flame. $500-700.

Unknown, ca. 1919. Most likely US made tobacco jar with a striker lighter on top. It has a pretty faceted glass with a bronze-like lid. 8" h. $350-450.

Unknown, ca. 1907. Early smoking related postcard. Notice the gentleman lighting his cigar with counter top cigar lighter

Unknown, ca. 1908. A sterling silver catalytic lighter. It used a special denatured alcohol that reacted with air and the tiny metal balls to produce a flame. 2.75" tall. $200-250.

Unknown, ca. 1920s to 1930s. An assortment of bakelite plastic lift arm lighters. All are American made. 2.5" tall. $40-70 each.

Unknown, ca. 1920. An Art Nouveau "Lady Diver" striker lighter. 7" l. $225-350.

Unknown, ca. 1920. A "Bull" striker lighter. Here is an example of a lighter by an unknown maker that has great collector value. $300-400.

Unknown, ca. 1933. USA. A standing elephant dressed in a suit. Looks very much like a Ronson. This political striker lighter represents the Republican Party. $500-900.

Unknown, ca. 1920. A "Indian Head" striker lighter. 5" high. $600-800.

Unknown ca. 1934. USA. Pretty enameled lighter and cigarette case with dog motif. $200-250.

Unknown, ca. 1920. USA. A "Bear" striker lighter. Wand located on his back. Flint strip along side of base 4" h. $250-350.

Unknown, ca. 1935. An American striker model with an advertisement for an iron and steel company in Minneapolis, Minnesota. $60-100.

Unknown, ca. 1935. A US made movie camera striker lighter. Marked "Hollywood." It has a copper finish. $200-400.

Unknown, ca. 1940. An unusual golf ball and golf club lighter. Moving the club lights the lighter. Front nut not original. Made in USA. $75-85.

Unknown, ca. 1938. A nice butterscotch colored bakelite striker lighter about 3" tall. The flint strip is on the bottom. $200-250.

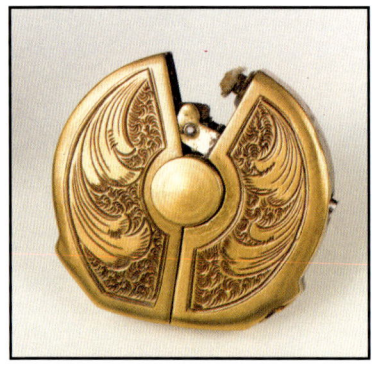
Unknown, ca. 1940. USA. An elaborately engraved, brass squeezer lighter. Unusual and rare piece. 1.7" x 1.7". $850-1,000.

Unknown, ca. 1939. An attractive lighter made for a woman's purse. It is made of enameled sterling silver and uses a simple wheel and wick. $125-175.

Unknown, ca. 1940. This lighter has an unmistakable Art Deco design. It is unmarked, but was probably made in the US. $40-60.

Unknown, ca. 1944. USA. Unusual trench art striker in the form of a frying pan. Note the striker, inserted below the handle and the striker surface forming the handle on the lid. Brass with steel striker. $500-750.

Unknown, ca. 1949. Metal lighter in the shape of the Johnson Wax Tower in Racine, Wisconsin. This well known building was designed by Frank Lloyd Wright. Lighter was produced for the building's dedication ceremony in 1950. Almost 6.5" tall h. $350-450.

Unknown, ca. 1950. "His Master's Breath" striker lighter. A take off of the RCA Nipper "His Master's Voice." Probably made in the US. $300-400.

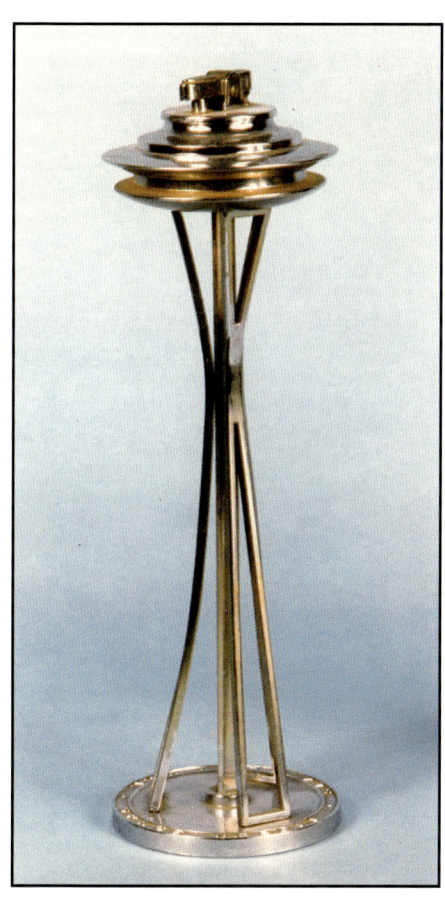
Unknown, ca. 1960. Table lighter in the shape of the space needle erected for the Seattle, Washington World's Fair. Made in Japan. 10" tall. $40-75.

Unknown, ca. 1953. This punch board had lighters as prizes. These items were declared illegal in most states since they were considered to be a form of gambling. The customer would pay a fee and for that fee would be allowed to punch out one of the holes. Each hole contained a tiny slip of paper telling you what you won or did not win. We are not sure what type of lighters were originally on the board. $75-100.

VALBERG, ca. 1914. Made in Rochester, New York, this four piece smoking set consisted of a striker lighter, ashtray, cigarette container, and base. Nicely decorated in a heavy silver plate. $400-600.

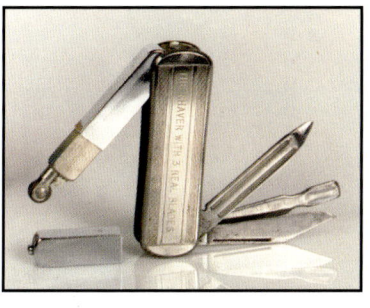
Unknown, ca. 1954. One side of this combination knife lighter says "Sunbeam Shavemaster, As Seen In Life Magazine" while the other says "The Shaver With 3 Real Blades." There is a simple lighter on one side and three blades on the other. Chrome plated metal. $125-175.

WALLACE, ca. 1950. A complete set of salt and pepper shakers and cigarette lighter in the Baroque style. $100-125.

WARNER, ca. 1940. The Warner lighter was made in Glendale, California. 2" tall. $45-75.

WESTON, ca. 1950. A full box of colorful Weston lighters. The box indicates that they also made the "Mighty Midgett," "Ball O Flint," and "Wonhand, Jr." lighters. $200-250 for all.

WEBSTER, ca. 1929. Two views of the lighter in closed position and with the windscreen and lid lifted. Made in the USA of sterling silver. $400-600.

WINDY, ca. 1942. A great advertising model with a Dr. Pepper soft drink ad. Manufactured by the Matawan Lighter Company of Matawan, New Jersey. $60-90.

WEBSTER, ca. 1929. You can lift one or both caps, allowing flexibility of a normal lighter and a windproof version. This lighter has a triple wheel flint mechanism. It is marked on the bottom "Webster sterling Lighter / Pat. Pending." $400-600.

WRIGHT, ca. 1911. The Wright lighter was made in New Jersey. This model is a nickel plated semi-automatic lighter. $125-175.

WRIGHT, ca 1911. Lighter fluid container and Dunhill box of snuff

 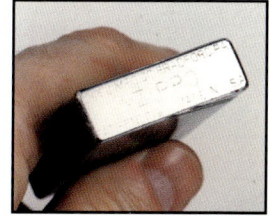

ZIPPO, ca. 1933. An extremely rare model produced in Zippo's first year of manufacture. Stands 3/8" higher than standard Zippo. Insert shown for size comparison. Hinge has three barrels. $12,000-16,000.

WRIGHT, ca. 1911. An early hand chased and engraved semi-automatic pocket lighter made of highly engraved sterling silver. $250-350.

YORK, ca. 1924. A bakelite lift arm pocket lighter. Made in the USA. $40-60.

ZIPPO, ca. 1933. A fabulous and extremely rare Art Deco Zippo table lighter. This model was produced in Zippo's first year of manufacture. It is tall, has an outside hinge and was a prototype produced in 1932/33. It is a one-of-a-kind experimental design which never actually went into production. It has red Scottie dog metalliques on both front and back, and a super Art Deco "fin" made of red "catalin," an early plastic, on its lid. The insert is 5/16 of an inch longer than those in standard 1934/35 outside hinge models. Height including base and red fin 3.25". Height without base is 3". Height of lighter itself from top of case to bottom of case is 2.5", and width is 1.5". Base measures 2" across. $40,000-60,000.

ZIPPO, ca. 1934. Outside hinge model with metallique ad for "Champion." Four barrel hinge. $4,000-6,000.

ZIPPO, ca. 1936. Engine-turned, prototype square model with attached emblem. $2,000-3,000.

ZIPPO, ca. 1935. Outside hinge with blue monogram metallique. $2,500-3,500.

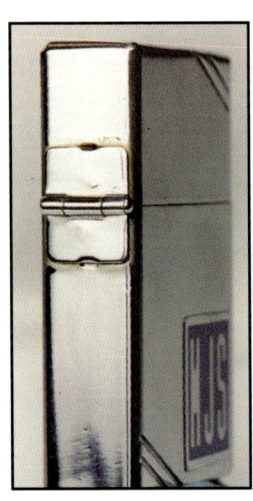

ZIPPO, 1936. Four barrel hinge metallique with ad for "Twin Disc" clutch company. Perfectly mint condition. $2,000-3,000.

ZIPPO, ca. 1935. Outside hinge. Extremely rare model with highly collectible Gulf Oil Co. metallique. $3,500-5,000.

ZIPPO, .1936. Square model with four barrel hinge and metallique advertisement for Joy Loaders. $1,800-2,800.

ZIPPO, 1936. A square four barrel model with highly collectible blue monogrammed metallique. $1,300-2,300.

ZIPPO, ca. 1936. without corner marks. Very rare. Dull finish possibly nickel. $2,500-3,500.

ZIPPO, 1936. A square four barrel model with mirror engraving for "Bradford Rotary Club." $1,200-1,600.

ZIPPO, 1936. A nice early square model with a "Sterling Gasoline" metallique logo. It sports the four barrel inside hinge. $1,800-2,800.

ZIPPO, 1936. A square four barrel prototype model. Experimental engraving without any corner slash marks. $1,400-1,800.

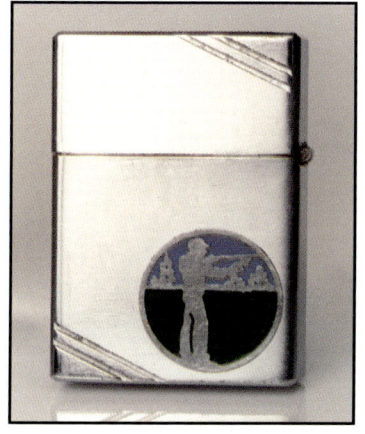

ZIPPO, ca. 1936. Two sided metalliques – a monogram on the front (not shown) and a shooter on reverse. This was an early outdoor theme much like their "Sport Series" line. $5,000-7,500.

ZIPPO, ca. 1936. Salesman's model in its original box. Shows the various advertising Zippo could offer their customers. Adorned with eleven metalliques including Kodak, Heinz 57, Gulf, Champion, Amoco, etc. Four barrel hinge with hump spring insert. $10,000-15,000.

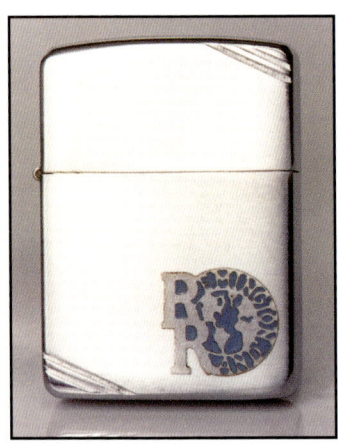

ZIPPO, ca. 1937. Metallique for Remington-Rand Company. Four barrel hinge. $1,500-2,200.

ZIPPO, ca. 1938. Art Deco designed by Belle Kogan, a famous Art Deco designer of the period. Belle Kogan Zippos are extremely rare and are among the most prized pieces in Zippo lighter collections. $4,000-7,500.

ZIPPO, ca. 1937. Metallique for Mobilgas with engraved monogram on reverse side. Four barrel hinge. $1,500-2,200.

ZIPPO, ca. 1938. Engraved "PCB" for Philo Blaisdell, father of Zippo founder, George Blaisdell. $1,500-2,500.

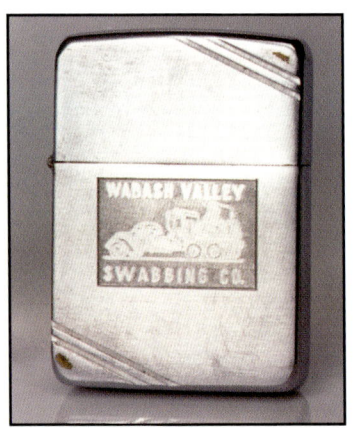

ZIPPO, ca. 1937. Reverse engraved for "Wabash Valley Swabbing Company." Four barrel hinge. $1,000-1,800.

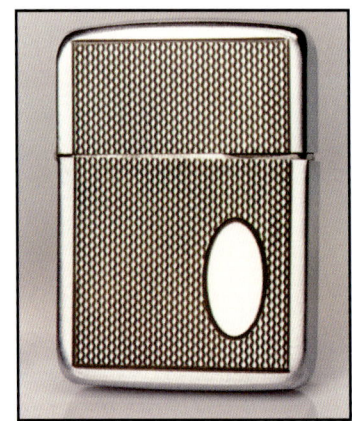

ZIPPO, ca. 1938. Prototype design. Experimental engraving which never went into production. $1,200-1,800.

ZIPPO, ca. 1938. Earliest known Town & Country "Lily Pond" with a four barrel hinge. Extremely rare to see Town & Country technique on a 1930s Zippo. $4,000-7,500.

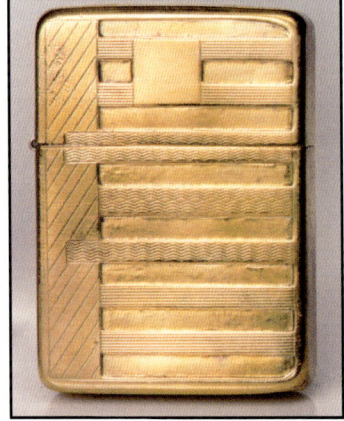

ZIPPO, ca. 1938-1939. Belle Kogan design "K-1" model. One of the most highly prized lighters among Zippo collectors. Extremely rare Art Deco model designed by this famous designer. $3,000-6,000.

ZIPPO, ca. 1938. Two-sided line drawings. Front has a bear, reverse has a spaniel. In-house experimental model. $1,500-2,000.

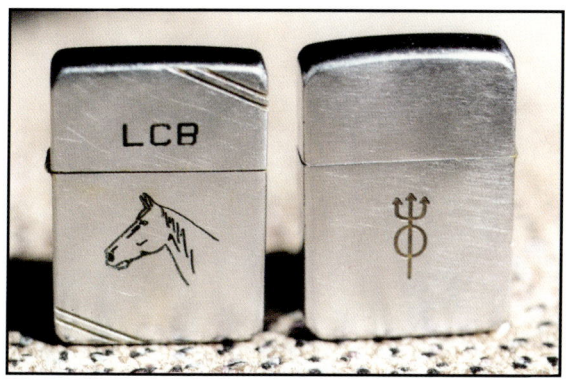 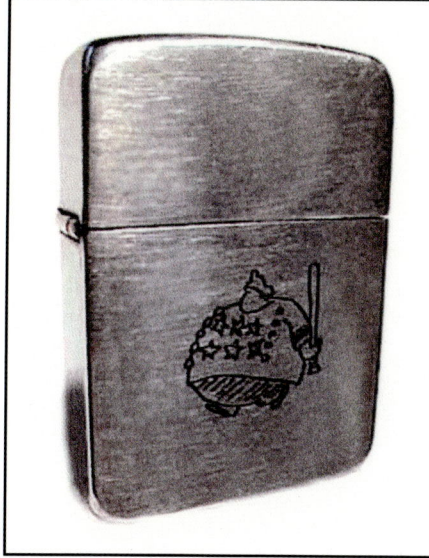

ZIPPO, 1938-1939. *Left:* Line drawing "Horse" from Zippo's first Sport Series of the 1930s. Four barrel hinge, plain insert. $500-700. *Right:* 1941. Line drawing of "Kelsey The Clown" of Saints & Sinner's Circus. On the reverse of this is the Keystone Cops logo. $600-1,000.

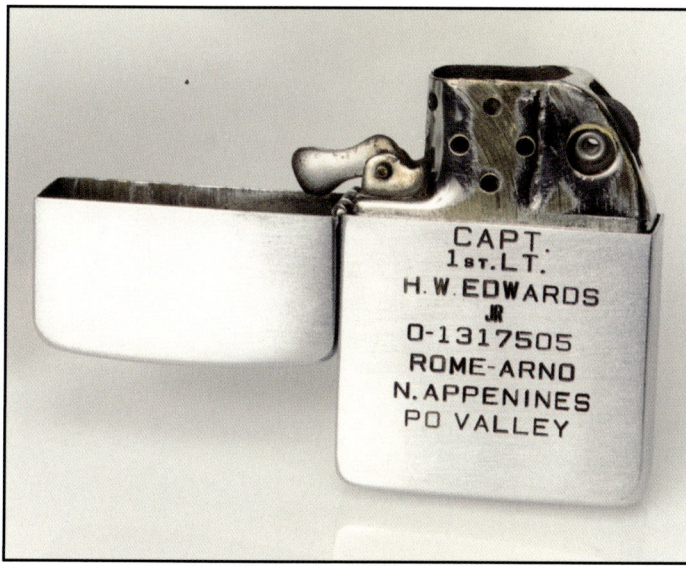 

ZIPPO, ca. 1938-1939. This lighter has a rare experimental insert and great military history. $2,500-3,500.

ZIPPO, ca. 1939. Unusual high-polished chrome finish. This model is most often seen with a brushed chrome finish, and rarely with this shiny finish. $800-1,200.

ZIPPO, ca. 1941. Unusual black enamel model four barrel hinge. $1,200-1,600.

ZIPPO, ca. 1939. A complete store wick display card. $125-175.

ZIPPO, ca. 1942. Steel model "Union 76 Oil Company." $400-700.

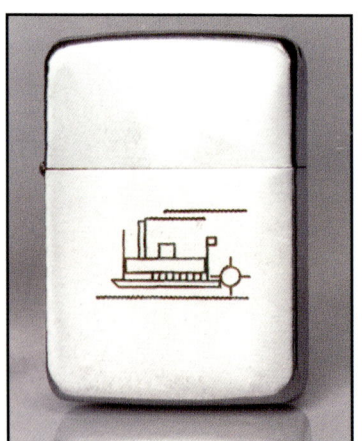

ZIPPO, ca. 1939-1940. Paddle boat line drawing. $600-1,000.

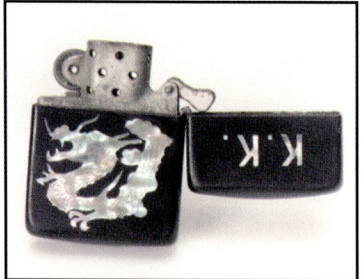
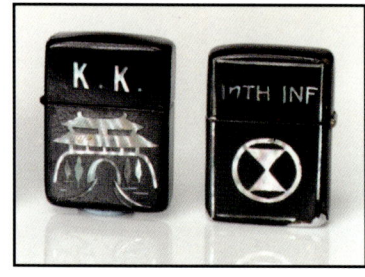

ZIPPO, ca. 1942 and 1946. Two abalone and black lacquer decorated lighters. The one marked "K.K." is a 1942 steel Zippo. The one marked "17th Infantry" is from 1946. $350-450.

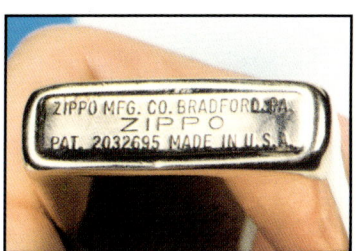

ZIPPO, ca. 1944. A 14kt gold Zippo lighter. This is an unusual piece since it has the indented, full patent marked bottom indicating that the case was made by Zippo rather than a jeweler. The 14kt mark is just below the five barrel hinge. Ignore the dark areas at the bottom of the lighter. They are not visible on the lighter and are merely a strange light reflection. $2,500-3,500.

ZIPPO, ca. 1945. Steel "In Memory Of Ernie Pyle," very famous war correspondent in the pre-television days killed at the end of WW II. One might say he was the Walter Cronkite of the 1940s. Only six hundred of these lighters were ever produced. $1,200-1,600.

ZIPPO, ca. 1946. A three barrel hinged, nickel silver model with a golfer. $300-500.

ZIPPO, ca. 1944. Interesting World War II era steel lighter with pins and coins depicting one person's travels. $200-250.

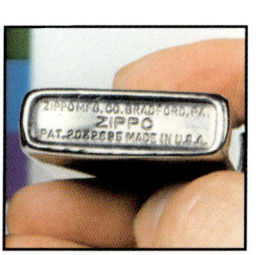

ZIPPO, ca. 1946. Town & Country "Lily Pond." Very rare "brushed chrome" finish. Patent 2032695 with three barrel hinge. $1,300-2,200.

ZIPPO, 1946. Prototype Town & Country "Gnome," and a Town & Country enamel "Horse." These were made between 1946 and 1952. $2,500-3,500 Gnome; $400-800 Horse.

ZIPPO, 1946. Early line drawings "Bowler" and "Golfer" both patented 2032695 with three barrel hinges. $300-400 each.

ZIPPO, ca. 1946-1947. Prototype design Patent 2032695. Three barrel hinge. Beautiful sunburst design on this in-house model. $1,500-1,750.

ZIPPO, ca. 1946. Patent 2032595 with three barrel hinge advertising Mohawk Gasoline. $200-250.

ZIPPO, 1946. *Left:* Early line drawing, "Bucking Bronco". Patent 2032695 with three barrel hinge. $400-600; *right:* 1946-1947 early line drawing, "Basketball Player." Patent 2032695 with three barrel hinge. $400-600.

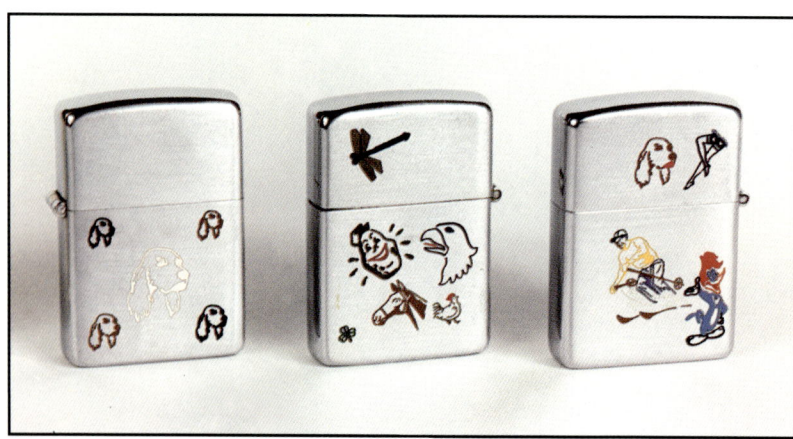

ZIPPO, 1946. The art department was busy trying out new designs on these lighters owned by a member of the Zippo design department. These were made between 1946 and 1952. $1,000 and up.

ZIPPO, 1946-1947. *Left:* Line drawing "Skeet Shooter." Patent 2032695 with three barrel hinge. $300-400; *right:* 1946 early "Test Sample Football Player" line drawing. Patent 2032695 with three barrel hinge. $500-700.

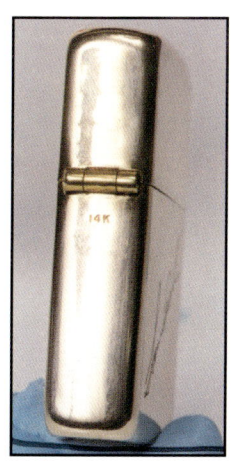

ZIPPO, ca. 1947. 14K gold Zippo engraved for "Sam Ash." Patent 2032695 with three barrel hinge. $2,500-3,000.

ZIPPO, ca. 1947. A "Barcroft 2nd" model table lighter in its box. This was produced from 1947 to 1950 and is seldom seen in the original box. 4" tall. $500-750.

ZIPPO, ca. 1947. Prototype red crackle model. Patent 2032695 with three barrel hinge. $1,400-2,000.

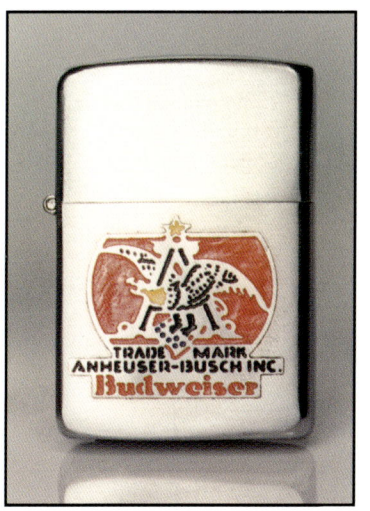

ZIPPO, ca. 1947. Town & Country Anheuser-Busch lighter. Patent 2032695 with three barrel hinge. $2,000-3,000.

ZIPPO, ca. 1947. Town & Country "Morning Glories" brushed finish. Rare model. $1,000-1,500.

ZIPPO, ca. 1949. A red full covered leather prototype without the usual gold border. $1,000-1,600.

ZIPPO, ca. 1949. Three barrel hinge with ad for Dupont. Rare Zippo produced only for one month in 1949. Unusual domed top and referred to as a "Rocket Top" Zippo by Linda Meabon at Zippo Mfg Co. Insert is nickel silver with sixteen hole chimney. $400-500.

ZIPPO, ca. 1949. Lucky Strikes cigarettes pack advertising model. Beautifully engraved. $400-700.

ZIPPO, ca. 1950. Town & Country model with multiple images. Beautiful example of Zippo T&C techniques of fifty years ago. $3,000-3,500.

ZIPPO, ca. 1949. Town & Country "Kitten" Patent 2032695 three barrel hinge. Extremely rare model. $1,400-2,400.

ZIPPO, ca. 1950. A two-sided advertising model which reads "Servel, The Gas Refrigerator," who were celebrating their 25th anniversary that year. $200-250.

ZIPPO, ca. 1950. This is a "Lady Bradford" table model Zippo with its original box. $300-400.

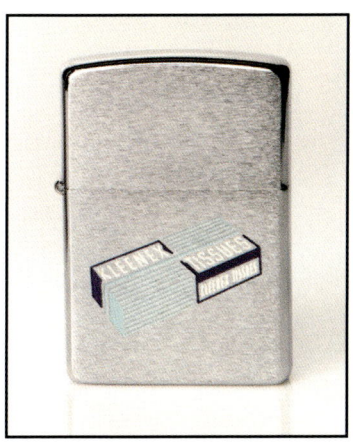

ZIPPO, ca. 1950. Canada Kleenex Tissue box with a three dimensional view. $200-250.

ZIPPO, ca. 1951. Town & Country "Drink Coca-Cola." At this time, this lighter is the only one known to exist. $4,000-7,500.

ZIPPO, ca. 1950. Canadian Academy Texaco. $150-250.

ZIPPO, ca. 1951. Town & Country - Presidential route of Dwight D Eisenhower's Presidential Campaign of 1952. $1,500-2,500.

ZIPPO, ca. 1950. "New Era Potato Chips Scientifically Processed." Beautiful Art Deco-style graphics. $200-250.

ZIPPO, ca. 1951. The Allstate "Guardsman" Tire. $200-250.

ZIPPO, ca. 1950. Zippo gift set with lighter, fluid, and flints. Lighter is a "Sailfish," one of Zippo's early "Sports Series" models. Value-$350-450.

ZIPPO, ca. 1951. Canadian Campbell's Tomato Soup, June 1951. It celebrates one million accident free hours. $300-350.

ZIPPO, ca. 1952. Two-sided with multiple engravings. This is a very rare one-of-a-kind Town & Country model commemorating Isabel Fuoco's retirement from the Zippo Company in 1952. The signatures all belong to Zippo employees. $1,500 and up.

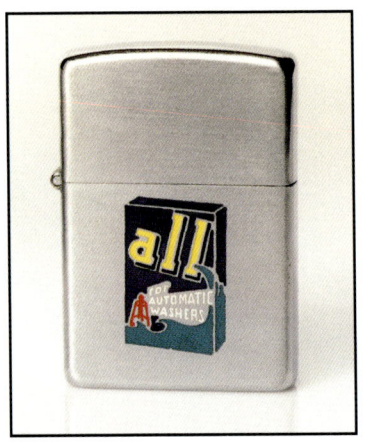

ZIPPO, ca. 1952. ALL clothes washing detergent "For Automatic Washers." With fantastic five color graphics. $300-350.

ZIPPO, ca. 1952. Blue leather ad for "Union 76" Gasoline. $400-700.

ZIPPO, ca. 1952. This sterling silver box is marked "Tiffany & Co., sterling, Italy" and holds a Zippo lighter. The ensemble was probably sold by Tiffany & Co. rather than Zippo. 2.75" tall. $400-500.

ZIPPO, ca. 1952. Extremely rare black gunmetal "Hunter with Dog," a one-of-a-kind "Loss-Proof" model. $2,500-4,500.

ZIPPO, 1952 and 1959. Manor House Coffee and Stokely Van Camp foods. $200-275.

ZIPPO, ca. 1952. A gold-filled lighter in a pretty green leather box advertising Cities Service gasoline. $200-250.

ZIPPO, ca. 1952-1953. sterling silver Town & Country sailboat. Extremely rare to see any Town & Country on a sterling silver lighter. $2,500-4,500.

ZIPPO, ca. 1952-1953. sterling silver extremely rare "Roy Rogers" one-of-a-kind model. $2,500-5,000.

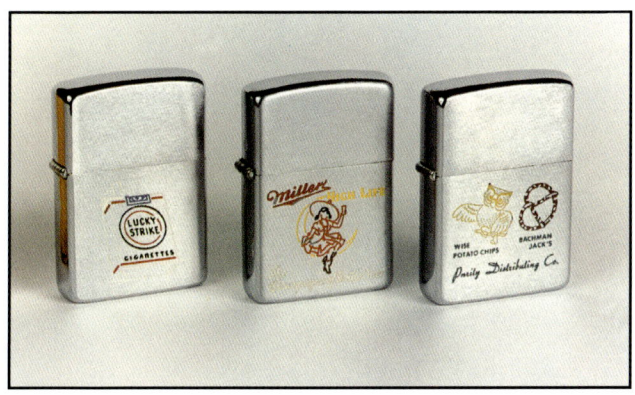

ZIPPO, 1953 and 1964. Three advertising lighters. The Miller Beer lighter is from 1953.

ZIPPO, 1953. Zippo displays with various lighters.

91

ZIPPO, ca. 1953 Personalized employee "Town & Country" model which never actually went into production. Extremely rare. Two-sided. $1,000 and up.

ZIPPO, ca. 1953. Roemer Insurance Co. with a nice old truck, "Safety First." $150-200.

ZIPPO, ca. 1953. Hastings Aero-type shrouded spark plugs advertising lighter. Beautiful graphics. $100-150.

ZIPPO, ca. 1953. Rohm & Haas 10W-30 Motor Oil. $150-200.

ZIPPO, 1953. "Wireman." Full patent 2517191 Patent Pending. Figural Wireman character smoking a cigarette. Value- $250-350.

ZIPPO, ca. 1953. Bell Aircraft Corporation. Two sided advertiser with "Bell Helicopters Fort Worth, Texas" on the front, and the Bell Aircraft logo on reverse side. $175-225.

ZIPPO, ca. 1953. Friden Adding Machines. It is colorfully engraved. $150-200.

ZIPPO, ca. 1954. Warner Brothers 25th Anniversary with Bugs Bunny. $400-500.

ZIPPO, ca. 1954. Howard Johnson's Restaurant and Motor Lodge. Two-sided advertiser with their well-known logo on reverse side. $175-225.

ZIPPO, ca. 1954. A 10kt gold-filled model with a Continental Casualty Insurance Company of Chicago logo. $150-200.

ZIPPO, ca. 1955. Beechcraft Super 18 Airplane on high polished chrome model Zippo. $175-225.

ZIPPO, ca. 1954. A Canadian Maxwell House Instant Coffee advertising lighter. Full-size model "Good To The Last Drop." Deeply engraved beautiful graphics. $200-250.

ZIPPO, ca. 1955. Smucker's Apple Butter Preserves and Jellies. $250-300.

ZIPPO, ca. 1955. Jantzen, The company is known for its swimwear. $125-175.

ZIPPO, ca. 1956. Town & Country winter scene, "Knob Lake". $1,000-1,400.

ZIPPO, ca. 1955. A Zippo promotional key ring. $80-100.

ZIPPO, ca. 1956. Town & Country "Bumblebee With Hive." $1,400-2,200.

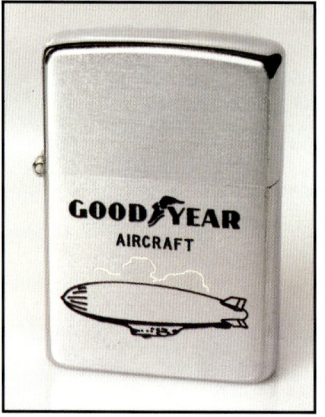

ZIPPO, ca. 1955. The Goodyear Blimp, the first company to use this unique method of "flying billboard" advertising. $200-350.

ZIPPO, ca. 1956. Town & Country extremely rare "Swordfish." $2,000-3,000.

ZIPPO, ca. 1956. Extremely rare blue anodized aluminum model produced for the Alcoa aluminum company in 1956. This is the only Alcoa aluminum Zippo model in blue, presently known to exist. There were, however, Alcoa blue aluminum models produced by both the "Park" and "Storm-Master" companies. $4,000-7,500.

ZIPPO, ca. 1956. Town & Country Airplane in flight. Extremely rare Town & Country model with beautiful enameling. $2,500-4,000.

ZIPPO, ca. 1957. Nicely engraved Dannon Yogurt advertiser. Beautiful graphics of a red Dannon container. $300-350.

ZIPPO, ca. 1957. A pair of lighters advertising Esso gasoline. $100-175 each.

ZIPPO, ca. 1957. Spring City Knitting Company's 50th Anniversary. $200-300.

ZIPPO, ca. 1957. A two-sided advertising model listing where Gleem Toothpaste was advertised at that time. $200-250.

ZIPPO, ca. 1957. GE Refrigerators & Freezers "Straight Line Design." $250-350.

ZIPPO, ca. 1957. Safety award for Beech-Nut baby foods and Life Savers roll candy. Beautifully engraved. $400-600.

ZIPPO, ca. 1958. Mrs. Inky slim model with a highly polished surface. $150-250.

ZIPPO, ca. 1957. Aluminum prototype slim model, dark blue color with steamboat engraved on one side. Reverse side has experimental pattern. Vintage aluminum Zippos are among the rarest with only a handful known to exist. $2,000-4,000.

ZIPPO, ca. 1958. Pretty Town & Country "Pheasant" model in mint condition. $500-750.

ZIPPO, ca. 1958. Permacel Tapes with beautiful colors. $200-300.

ZIPPO, ca. 1958. Town & Country "Mallard" model with a flying mallard, in mint condition. $500-750.

ZIPPO, ca. 1958. Procter & Gamble Spotlight Award. Procter & Gamble sponsored many of the daytime television shows in USA. $150-250.

ZIPPO, ca. 1958. A Town & Country "Trout " model in mint condition. This model is engraved with a trout rising to take a fly. $500-750.

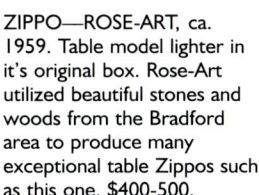

ZIPPO, ca. 1958. Hotel Cadillac with wonderful early engraving. $150-200.

ZIPPO, ca. 1958. A Town & Country model in mint condition with a Bell Aircraft Industries logo. This is one of a group of T & C models made for Bell Aircraft. $650-950.

ZIPPO, ca. 1958. A "Lossproof" Hunter motif model with its lanyard and special box. $300-350.

ZIPPO—ROSE-ART, ca. 1959. Table model lighter in it's original box. Rose-Art utilized beautiful stones and woods from the Bradford area to produce many exceptional table Zippos such as this one. $400-500.

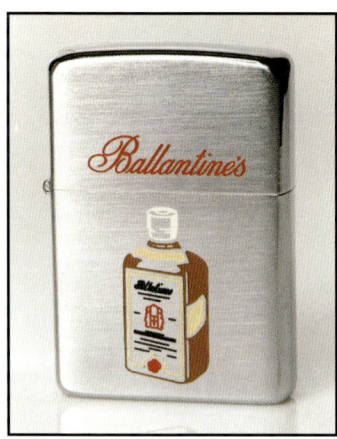

ZIPPO, ca. 1959. Ballantine's Scotch Whiskey. The reverse side, in Spanish, says "Selecciones del Reader's Digest" Magazine. It is beautifully engraved on both top and bottom. $200-300.

ZIPPO, ca. 1960. Canadian Mitchum Deodorant Roll-On with beautiful graphics. $150-250.

ZIPPO, 1959. RCA Whirlpool Home Appliances lighters (pocket and table) Wonderful graphics. $200-300 each.

ZIPPO, ca. 1960. Tri-City Ice Co. of Freeport, Texas. $200-300.

ZIPPO, 1959. A wonderful image on an advertising lighter showing the Sylvania TV. $250-350.

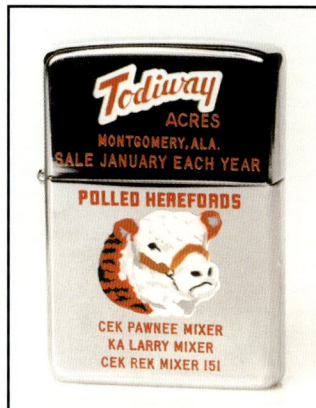

ZIPPO, ca. 1960. Town & Country advertising Todiway Acres, Polled Herefords. $900-1300.

ZIPPO, 1960 & 1970. Enjoy Coca Cola In Bottles and Holiday Inn Zippo salesman's advertising lighters. The Coca Cola lighter is from 1960. $250-350.

ZIPPO, ca. 1960. Town & Country Swan on a slim model. $1000-1500.

ZIPPO, ca. 1961. A 10kt gold-filled slim model with a Pepsi Cola advertisement. $200-250.

ZIPPO, ca. 1960. White House with "Compliments President Dwight Eisenhower." $400-600.

ZIPPO, ca. 1961. Pacific States Box & Basket Co. from Los Angeles, California. $150-250.

ZIPPO, ca. 1961. Very rare "Town & Country" model with a flower on one side and an advertisement for the Grand Hotel Alabama on the reverse. $2000-3000.

ZIPPO, ca. 1961. Golden Flake was an early breakfast cereal. $150-250.

ZIPPO, ca. 1961. Owens-Illinois Paper Products Division. $150-250.

ZIPPO, ca. 1962. McDonald's Hamburgers Drive-In, Mason City, Iowa. Beautiful early engraved McDonalds Restaurant. Value if excellent, $500-900.

ZIPPO, ca. 1961. Slim model of Maxwell House Coffee cup "Good To The Last Drop". $150-200.

ZIPPO, ca. 1962. Steelcase Office Furniture Company. Beautiful multi-colored graphics. $150-250.

ZIPPO, ca. 1962. A nice old Saab Automobile. $150-250.

ZIPPO, ca. 1962. A Town & Country model with an advertisement for Colorado National Sugar. $600-900.

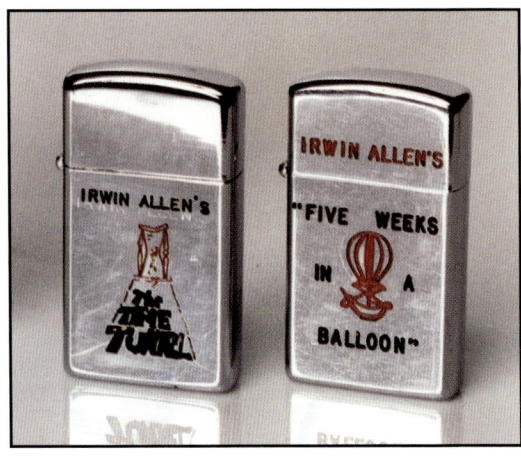

ZIPPO, ca. 1962. Two great advertisement lighters. Each advertises a different Irwin Allen Broadway show – *Five Weeks in a Balloon* and *The Time Tunnel*. $150-250 each.

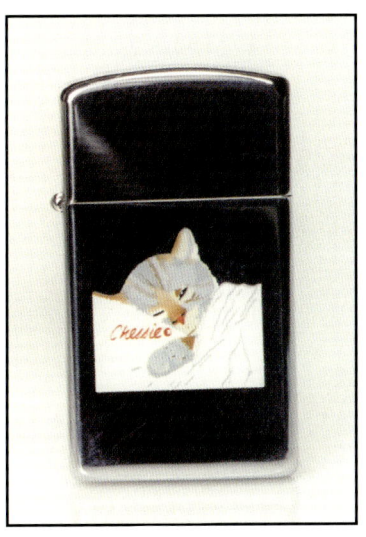
ZIPPO, ca. 1963. Town & Country style, Chessie The Cat, mascot of the Chesapeake Railroad. Slim model. $300-500.

ZIPPO, ca. 1963. Peter Pan Peanut Butter. Wonderful graphics on this much sought after advertiser. $300-500.

ZIPPO, ca. 1963. Town & Country with nice graphic of a cow, advertising Hereford Farms Florida. $900-1,100.

ZIPPO, ca. 1963. "From The White House Staff," on one side and "Christmas 1963". $300-500.

ZIPPO, ca. 1964. An unusual slim sterling silver model with a cowboy on it. $200-250.

ZIPPO, ca. 1964. Terminix, the exterminating company. Wonderful graphics on this advertiser. $100-150.

ZIPPO, ca. 1965. A Town & Country model with an advertisement for *The Shreveport Times* newspaper. $250-350.

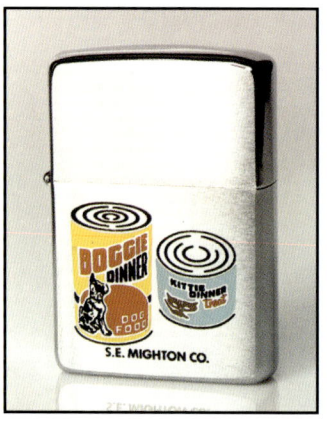

ZIPPO, ca. 1964. S. E. Mighton Co. advertising Doggie Dinner dog food & Kittie Dinner Treat. $300-500.

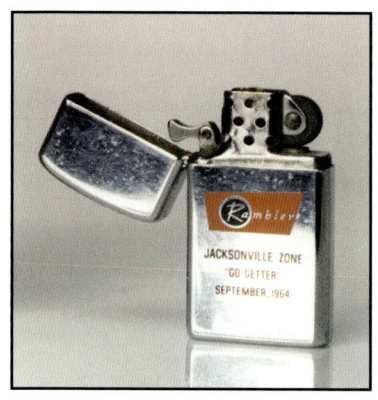

ZIPPO, ca. 1965. A slim sized lighter with a Rambler Automobile advertisement from 1965. $100-150.

ZIPPO, ca. 1965. Town & Country process C. G. Bark Eagle. $500-700.

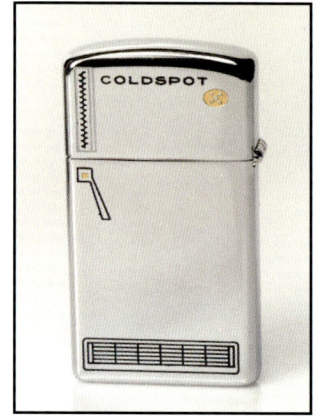

ZIPPO, ca. 1965. A Coldspot Refrigerator on a highly polished slim model. $250-350.

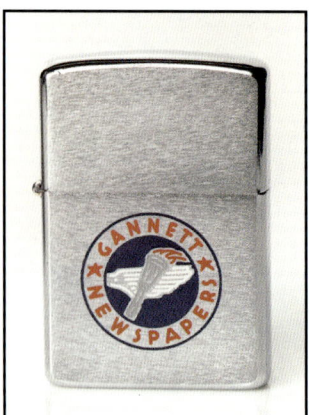

ZIPPO, ca. 1965. Gannett Newspapers advertising. $75-150.

ZIPPO, ca. 1965. Advertising for Bell & Howell Specialist film projector. This early reel to reel film projector was considered one of the finest of this time. $200-250.

ZIPPO, ca. 1967. Very rare Zippo 29¢ fuel container. $200-300.

ZIPPO, ca. 1966. An image advertising the United States Pavilion at Expo 67, on a standard size lighter. $100-150.

ZIPPO, ca. 1967. Slim model with a Ford Motors Dealer advertising. $60-80.

ZIPPO, ca. 1966. Towmotor Fork Lift Trucks advertisement. $100-150.

ZIPPO, ca. 1967. A California Seals Hockey Team lighter. $75-100.

ZIPPO, ca. 1967. Jack Frost Cane Sugar with a well known advertising character who has adorned many sugar products and continues to do so today. $250-400.

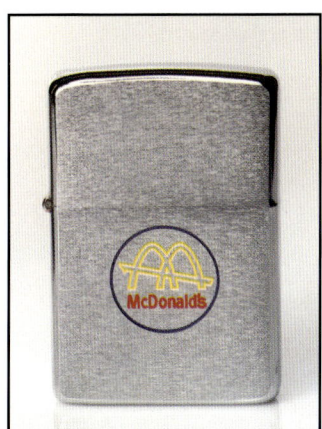
ZIPPO, ca. 1967. McDonald's Hamburgers. Early and extremely rare Zippo. $400-800.

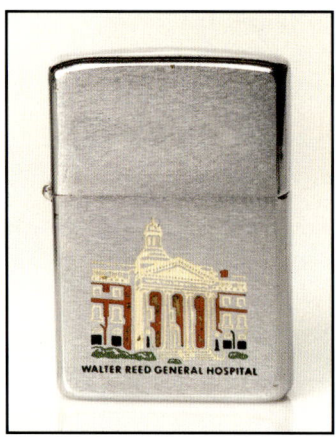
ZIPPO, ca. 1967. Walter Reed General Hospital. It is very unusual to see a hospital promoting itself on a Zippo, or any other lighter. $150-250.

ZIPPO, ca. 1968. Burger Chef Hamburgers on a highly polished chrome slim model. $150-250.

ZIPPO, ca. 1968. Curaçao Synagogue. Large, yellow engraved building. Very rare to see a synagogue on a Zippo. Curaçao Synagogue was the first synagogue to be built in all of the Americas. $150-200.

ZIPPO, ca. 1969. Extremely rare Town & Country transitional Astrology prototype with fabulous colors. $3,000-5,000.

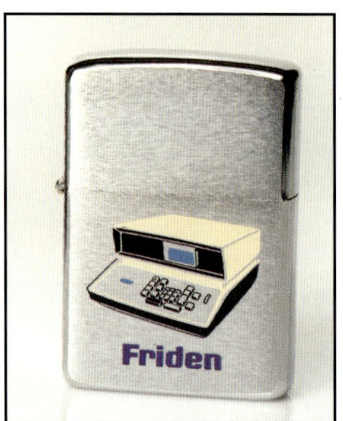
ZIPPO, ca. 1968. Friden Office Machine. The adding machine has a great early computer look. $200-250.

ZIPPO, ca. 1969. Extremely rare Town & Country transitional Zippo made for NASA. Features astronauts Armstrong, Collins, and Aldrin. Beautiful colors. $3,000-5,000.

ZIPPO, ca. 1968. Engraved with advertisement for Roman Meal bread. $100-150.

ZIPPO, ca. 1969. Extremely rare Town & Country transitional Zippo with image of George Blaisdell, founder of Zippo. Deeply etched with multi-colored enamels. $3,000-5,000.

ZIPPO, ca. 1969. Town & Country transitional Zippo. Slim Lily Pond, also referred to as Lily Pads. Most often seen on a regular size model. $500-750.

ZIPPO, ca. 1969. Prototype Aquarius transitional Town & Country Zippo. Beautiful colors on this extremely rare Zippo. $3,000-5,000.

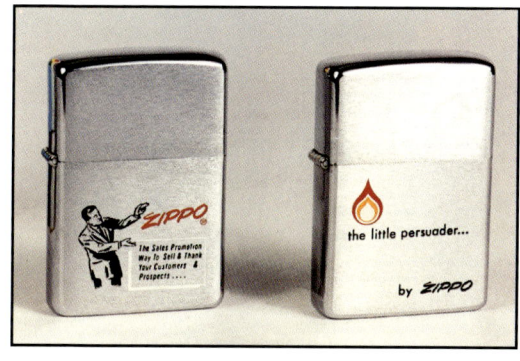

ZIPPO, ca. 1970 and 1973. Two salesman's Zippos advertising the Zippo company. $175-225 each.

ZIPPO, 1969. A scarce all copper lighter made for the Kennecott Copper Company. $400-700.

ZIPPO, ca. 1971. U-Haul Truck Rental Company. Very colorful graphics. $100-150.

ZIPPO, ca. 1969. Prototype with Signs of the Zodiac. Town & Country transitional lighter with many beautiful colors. Extremely rare model. $3,000-5,000.

ZIPPO, ca. 1971. A Disneyland image engraved "Walt Disney Productions." $300-400.

ZIPPO, ca. 1976. A Zippo knife with an advertisement for Kendall Oil's 95th anniversary and America's Bicentennial commemoration. $100-125.

ZIPPO, ca. 1972. M&M Body Repair, Abilene, Texas. Note the bikini-clad woman in high heels. $100-150.

ZIPPO, ca. 1977. Master Charge, the original name of today's "Mastercard." Raytown Bank, Missouri. Slim size. $75-125.

ZIPPO, ca. 1974. Nice advertisement for the Yellow Pages phone directory. $75-125.

ZIPPO, ca. 1977. An extremely rare and highly desirable Raggedy Ann model lighter. $800-1200.

ZIPPO, ca. 1976. Bumble Bee Seafood, well known tuna and seafood distributor. $200-250.

ZIPPO, ca. 1972. Swiss Miss, manufacturer of the much loved hot chocolate drink. $150-250.

ZIPPO, ca. 1977. Extremely rare Schooner – genuine scrimshaw engraving on bone, not acrylic as later models were. Beautiful water scene with lighthouse on lid, and large boat on the bottom. Scrimshaw "chips" on brass lighter. $1,000-1,400.

ZIPPO, ca. 1980. Unusual and very rare Mighty Mouse graphics. We have seen a few other Mighty Mouse Zippos also produced during this time. $700-900.

ZIPPO, ca. 1977. A pair of lighters, one slim and one standard size model with an image of the Bricklin Motor Car. *Left* (standard size): $225-300; *right* (slim): $150-200.

ZIPPO, ca. 1980. A Pittsburgh Pirates baseball team lighter. $75-100.

ZIPPO, ca. 1977. Alaska Airlines, "The sky's the limit." $125-175.

ZIPPO, ca. 1980s. This is a difficult-to-find lighter that was available only in Bradford, PA to family and friends of the employees of the Zippo Company. Engraved "Congratulations Graduate." $75-100.

ZIPPO, ca. 1979. A Zippo retractable tape measure with a great advertisement for Holsum Bread. $50-75.

ZIPPO, ca. 1981. A US Naval Sealift Command Zippo. $150-200.

ZIPPO, ca. 1984. This is a "Handilite" table lighter advertising Kendall Motor Oil. $200-250.

ZIPPO, ca. 1981. Campbell's Condensed Soup. A highly prized Zippo, immortalizing Andy Warhol's famous painting of a soup can. $400-600.

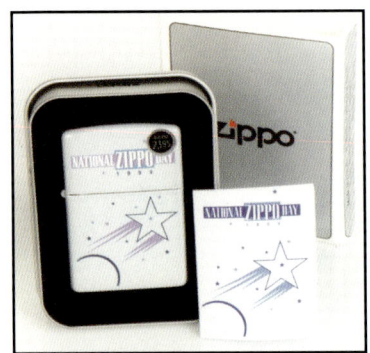
ZIPPO, ca. 1999. This lighter was available at the official Zippo Swap meet in Bradford, PA and is a limited edition piece. $100-150.

ZIPPO, ca. 1982. "Lucky Strikes Again" is the wording on this cigarette advertising model. $100-150.

ZIPPO, 2001. sterling silver lighter with various gold emblems. Someone adorned their favorite lighter with fraternity and lodge pins along with Mickey Mouse and the RCA dog, Nipper. $100-150.

ZIPPO, ca. 1990s. Original Simon M. Lytton "Eagle" engraved Zippo.

ZIPPO, ca. 1990s Original Simon M. Lytton "English Scroll" engraving on Zippo.

ZIPPO, ca. 1990s. Original Simon M. Lytton "Nude" engraved Zippo.

ZIPPO, ca. 1990s Original S. Lytton "Eastern Dragon" engraving on Zippo.

ZIPPO, ca. 1990s. Original Simon M. Lytton "Leaf" engraved Zippo.

ZIPPO, ca. 1990s. Original S. Lytton "Western Dragon" engraving on Zippo.

ZIPPO, ca. 1990s Original Simon M. Lytton "Owl" engraved Zippo.

ZIPPO, ca. 1990s. Photo of artist Simon M Lytton while engraving.

ZIPPO, ca. 1990s. Five Simon Lytton engraved Zippos. Shown here are Frog & Leaf design, Moths & Leaf design, Dragons, Flower design, Leaping Trout, and English Scroll. Photo by Weyer of Toldeo, Ohio.

ZIPPO, ca. 1990s. Four Simon M Lytton engraved Zippos from the 1990s. Shown are the Dragon, Smoke Lady, and Koi Tattoo designs.

ZIPPO, 2000. Beautiful one-of-a-kind "Statue Of Liberty" by Claudio Mazzi, renowned Italian artist. Purchased at Zippo Swap Meet 2000 fund raising auction for $1,700. Signed "Mazzi 2000" in white below the hinge. Mr. Mazzi, with his wonderful air brushing techniques has created and continues to create many fabulous designs on Zippo lighters.

ZIPPO, ca. 1990s. Five Simon Lytton engraved Zippos. Shown here are American Scroll with beaded background, Koi Carp Tattoo design, Smoke Lady, and Art Deco Flowers.

France

ABDULLA, ca. 1931. A short silver-plated model with a lacquer design of two golden birds. 1.5" x 1.6". $1,750-2,000.

ABDULLA, ca. 1929. France. sterling silver semi-automatic model with 18kt gold stripes. Note that the plunger can be prevented from activating by tightening the screw. 2" x 1.6". $1,750-2,000.

ABDULTRA, ca. 1936. France. A chrome plated watch lighter with a very unusual digital display. 2" x 1.6".

ABDULLA, ca. 1929. France. A 9kt gold sleeve over brass. 2" x 1.6". $750-1,000.

AUTOLUX, ca. 1952. Two French chrome plated lighters made by Autolux. $60-80 each.

ABDULLA, ca. 1929. France. A rare sterling silver model with a black lacquered bi-plane on one side and enamel work on the other side. 2" x 1.6". $2,500-3,000.

ABDULLA, ca. 1932. A short chrome plated model with an engine-turned design. 1.5" x 1.6". $150-225.

AUTOLUX, ca. 1952. A French nickel plated and leather covered lighter made by Autolux. $50-70.

BICHROMATE, ca. 1880. French Catalytic lighter. The larger bottle was filled with a potassium bichromate solution and also had two carbon electrodes descending into the bottle. When the spring loaded rod was pushed down, a zinc electrode entered the solution and an electric charge was created. The smaller bottle was filled with gasoline. When the rod was pushed down, it opened the cover of the gasoline infused wick. The electricity caused platinum wires to glow and ignite the wick.

AUTOLUX, ca. 1955. Three French Autolux automatic lighters. 2" tall. $50-60 each.

BOUCHERON, ca. 1942. Elegant and rare French made 18K gold lighter with an Imco-type mechanism by Boucheron. $3,000-4,000.

BBR, ca. 1928. An interesting French chrome lift arm lighter advertising Spido oil additive. There is a tax stamp on the one side. $125-175.

CARTIER, ca. 1929. A pair of superb watch lighters in 18kt gold and black enamel. The one on the left has a ring for a chain. These are two elegant pieces of Art Deco design art. 1.8" x 1.3". $20,000-25,000 each.

CARTIER, ca. 1929. 18kt gold with a nice diamond checkerboard engine turning. It has rounded edges and a protruding snuffer cap that entirely covers the flint wheel mechanism. It has an unusual monogram on the bottom and engraved on the side of the lid is the date Oct. 21, 1929 (eight days before the stock market crash). Marked "Cartier Paris." 1.9" x 1.4". $4,000-6,000.

CARTIER, ca. 1930. 18kt gold and black enamel with an engine turned design including diamonds. It has a round protruding snuffer cap. Marked "Cartier Paris." 1.7" x 1.4". $7,500-8,500.

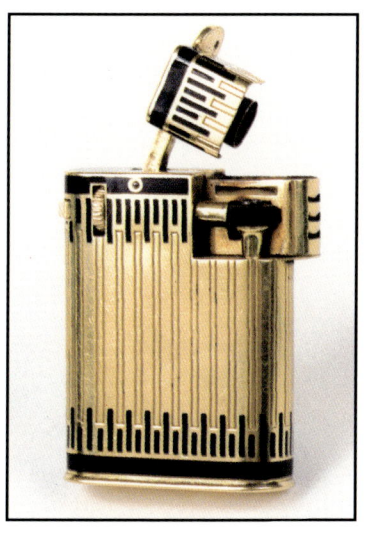

CARTIER, ca. 1930. An 18kt gold engine turned design with black enamel detail. It has rounded edges and a protruding snuffer cap that entirely covers the flint wheel mechanism. This is a very rare large size with a wonderful windguard. Marked "Cartier Paris". 2" x 1.6". $12,500-15,000.

CARTIER, ca. 1930. 18kt gold with a great Art Deco design and a central monogram. It has a rounded edge, protruding snuffer cap that entirely covers the flint wheel mechanism. Marked "Cartier Paris." 1.6" x 1.4". $7,500-8,500.

CARTIER, ca. 1930. 18kt gold with black enamel detail and diamond trim. It has rounded edges and a protruding snuffer cap that entirely covers the flint wheel mechanism. Marked "Cartier Paris." 1.7" x 1.4". $5,000-6,000.

CARTIER, ca. 1930. Table lighter with a mirror backed glass base. The lighter is sterling silver with black enamel. It has a rounded edge, protruding snuffer cap that entirely covers the flint wheel mechanism. It appears to be a pocket lighter inserted permanently into the base. Marked "Cartier Paris Depose." 2.8" x 2.1". $3,000-4,000.

 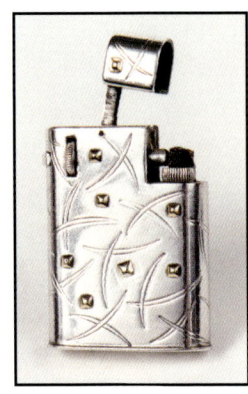 

CARTIER-Type, ca. 1930. Three lighters with a similar size and shape that are believed to have been sold by Cartier, but not marked as such. Apparently produced by the same atelier that Cartier Paris used, as they are visually identical to Cartiers. They share the rectangular shape and rounded edge, protruding snuffer cap that entirely covers the flint wheel mechanism. *Left:* sterling silver with an elaborate flower engraved on both sides with an elaborate attached gold pistil in the center. Monogrammed on the bottom. Similar hallmark and engraved with a work number. *Center:* sterling silver with Art Deco engraving and sixteen gold rivets in a random pattern. 1.8" x 1.5". Similar hallmark and engraved with a work number. 1.9" x 1.5". *Right:* sterling sliver with a very Art Deco engine turned design and engraving. It bears a French hallmark for silver and has a work number engraved. 1.8" x 1.3". $1,500-3,500 each.

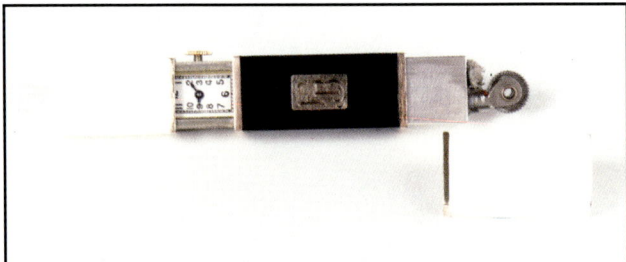

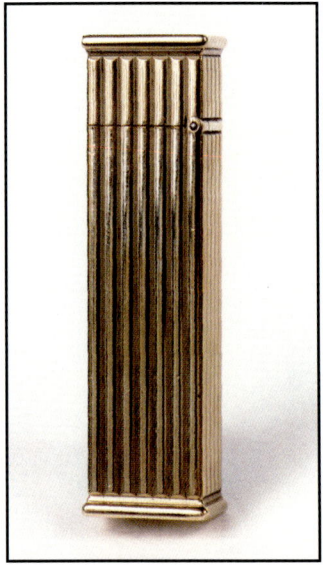

CARTIER-Type, ca. 1930. A watch lighter that is believed to have been sold by Cartier, but not marked as such. Apparently produced by the same atelier that Cartier Paris used, as it is visually identical to a Cartier. Small rectangular purse watch in a combination of silver, gold, and black and white enamel. One end of the piece slides up (but not off) to reveal the small watch. The other end slides off to reveal a simple flint and wheel mechanism. There is an elaborate monogram on one side of the lighter on a gold plaque. 3" x 0.3". $1,500-2,000.

CARTIER, ca. 1932. 9kt gold tall ribbed, with full lift top and exposed thumb wheel for operating the flint wheel. Marked "Cartier London." 2.6" x .7". $2,500-3,500.

CARTIER, ca. 1932. A watch lighter in sterling silver and black enamel. Extremely rare. This level of workmanship was generally reserved for gold lighters 1.8" x 1.3". $15,000-17,500.

CARTIER, ca. 1932. sterling silver and 18kt gold with black enamel. It comes in its original box. $3,000-4,000.

CARTIER, ca. 1932. An 18kt gold and black enamel model with a watch. This has a beautiful Art Deco design with an enameled bezel. 1.75" h. x 1.25" w. $20,000-25,000.

CARTIER, ca. 1935. 18kt gold with a very fine engine turning with black lacquer trim top and bottom. It has a rectangular, slightly protruding snuffer cap that entirely covers the flint wheel mechanism. Marked "Cartier Paris." 1.9" x 1.3". $6,000-7,000.

CARTIER, ca. 1934. Dunhill style watch lighter in 18kt gold Florentine pattern with diamonds and rubies on the body of the lighter serving as numerals. There is a ruby on the flint screw as well. We believe this to have been manufactured by Dunhill. The curved lift arm is the Dunhill style but unmarked. We have seen pieces marked both Dunhill and Cartier on the base, but the lift arm remains unmarked. 2.1" x 1.1". $10,000-12,500.

CARTIER, ca. 1943. A pair of 18kt gold lighters that are the same size and shape, each with different engine turning. Both have oval, protruding snuffer caps that entirely cover the flint wheel mechanism. Both have set diamonds, one with a twelve-faceted pavé setting and the other with full cut fifty-six-faceted diamonds. The one on the left has pave-set diamonds and is marked "Cartier France 14kts." 1.5" x 1.4". $5,000-7,500. The one on the right is marked "Cartier France 5ca (total weight of the diamonds), 18kts, Cartier Paris." 1.5" x 1.4". $7,500-10,000.

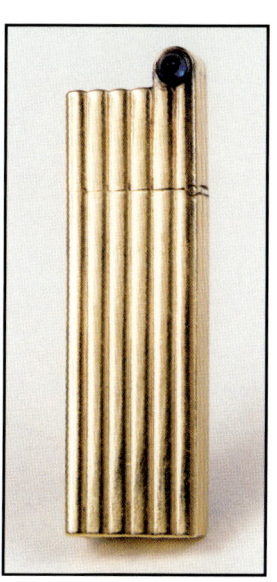

CARTIER, ca. 1935. 18kt gold tall ribbed automatic with a sapphire cabochon thumb piece. Marked "Cartier Paris." 2.8" x .8". $4,500-5,500.

CARTIER, ca. 1949. France. sterling silver with a vertical ribbing and rectangular, protruding snuffer cap that entirely covers the flint wheel mechanism. Marked "Cartier Paris Depose." 1.8" x 1.4". $750-1,000.

CARTIER, ca. 1950. Four lighters with a similar size and shape. All are marked "Cartier Paris." They date from the late 1940s to the mid 1950s. All have rectangular, protruding snuffer caps that entirely cover the flint wheel mechanism. *Left:* 18k vertically ribbed with green gemstone thumb piece. *Left center:* sterling covered entirely in black enamel except for the sterling screws and a gold initial plate attached to the front of the snuffer cap. *Right center:* Same as the lighter on the left with broader ribs and a monogram in lieu of the thumb piece. *Right:* Same as the lighter in the left center, except the front of the snuffer cap is also enameled and the top of the snuffer has a small sapphire immaculately carved in the shape of a jaguar's head. 1.6" x 1.4".

CLARODAN, ca. 1933. Although marked "Hand Made," this is actually a machine made silver plated lighter with a top mounted thumb roller. It was made when mass production was out of favor in Great Britain; hence the misleading mark. $100-150.

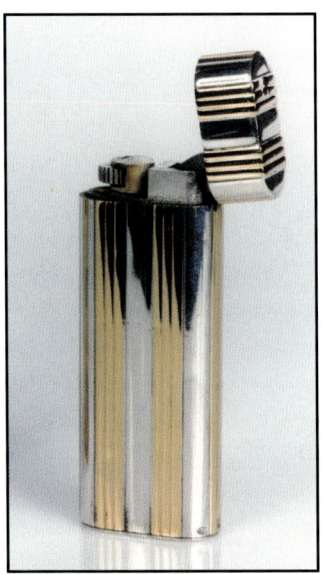

CARTIER, ca. 1974. France. A sterling silver lighter with 18kt gold vertical bands. It uses butane. $750-950.

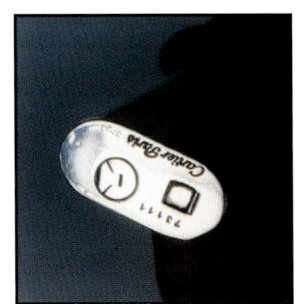

DRAGO & FLAMIDOR, ca. 1933. Two French made lighters. The one on the left is a Drago "Type Special" lift arm with roller on top. It bears the insignia of the French Foreign Legion. The one on the right is a Flamidor manual lift arm, chrome plated lighter with an unusually large wheel. *Left:* $400-500; *right:* $ 40-60.

CHIC, ca. 1928. Two beautiful, French, sterling silver, pull-out, automatic spark lighters. They exhibit the finest enameling we've seen. $2,500-3,000 each

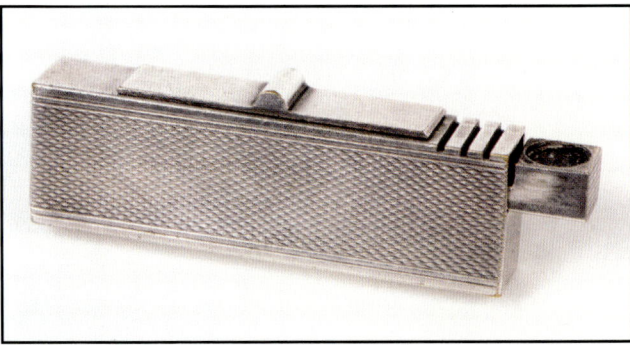

DUCAT, ca. 1939. France (surmised). A silver plate chemical slide action lighter. When slide is opened it reveals the small screened area that would heat up upon exposure to the air. 2.5" x .8". $250-300.

S. T. DUPONT, 1942. A French made silver plated table "Jeroboam" lighter. A nice feature is the second wick on a removable wand. $1,500-2,000.

S. T. DUPONT, ca. 1942 France. "Jeroboam." A French made gold plated petrol model with black lacquer highlights. This model uses petrol and has second wick on a removable wand. $1,500-2,000.

S. T. DUPONT, ca. 1942. This is a pocket lighter reminiscent of the Jeroboam table lighter described above. Notice the lift arm and horizontal rod in place of the typical lift arm with wheel. This one is gold plated with Namiki lacquer detail. On the side edge it bears the distinct red signature of its Japanese artisan. It is not marked Dupont because this was a line that Dupont brought to market for the more exclusive smoker. It is branded on the bottom "Le Sultan, Paris, Brevete tous pays" (patented all countries). $750-1,000.

S. T. DUPONT, ca. 1945. Two seemingly identical table lighters except for the finish. The one on the left is the only one known to re-flint by lifting the rear arm (shown in the up position). The one on the right is more typical with a small flint screw on the bottom and a long spring that runs to the top of the lighter. The lighter pictured on the left is said to have come from the estate of a longtime Dupont craftsman. Also of note is the removable "pass-around" accessory. To the right of the lift arm, in both examples, notice the tubular knob. Pull on this and a tube detaches. Turn the tube upside down and you have a wick and fuel system. Hold that to the lighted lighter; the tube ignites and you can pass it around without having pass the entire lighter to accomplish that purpose. Both versions are equally made of the highest quality. Both lighters are marked on the bottom with the script version of "S. T. Dupont," "Jeroboam." The model on the left bears "Paris Brevete tous pays" while the more common model on the right bears the anglicized "PARIS MADE IN FRANCE. *Left*: $1,200-1,600; *right*: $750-$1,100.

S. T. DUPONT, ca. 1965. France. Standard Large butane Windsor model. Black enamel with a gold plated trim. 1.5" h. $175-225.

S. T. DUPONT, ca. 1970. France. Line 1 model butane lighter in a large sized green enamel with a gold plated trim. 2" h. $200-250.

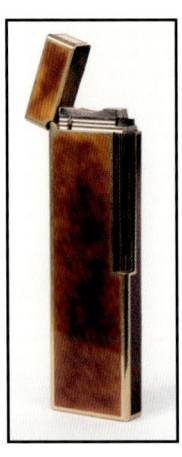

S. T. DUPONT, ca. 1975. France. Line I table model lighter in a tortoise enamel with gold plated trim. 6" l. $300-350.

S. T. DUPONT, ca. 1980. France. Line 2 large butane model with a silver plated trim with alternating engine turned panels. 1.5" h. $150-200.

ERLAC, ca. 1942. Three Erlac semi-automatic lighters. The top lifts when you press the flint wheel in firmly. It has an unusual and tricky mechanism for flint and fuel. To release the flint holder you must press the thumb wheel in and up at an angle. *Left:* sterling silver ribbed version, marked on bottom "Dunhill PARIS" "Licence (not license) Erlac" and "Brevet No. 518687." *Center:* Silver plated model with wide ribs, marked on bottom "Erlac" (in script) and "Breveté S. G.D.G. *Right:* A brass ribbed engine-turned version, marked identically to the one above. All are 1.9" x 1.4". *Left:* $1,400-1,800; *center and right:* $400-600 each.

ÉLYSÉE, ca. 1942. sterling silver lighter made in Paris, France with the same high quality and craftsmanship as the Cartier lighters of that time. $400-500.

FLAMIDOR, ca. 1938. Two fuel cans in nickel plated brass. The one on the left is marked "Flamidor Paris." On the right is a smaller one marked on the bottom "DRPA" and "D.R.G.M." $75-125 each

EPSOM, ca. 1924. A Paris, France made lighter in silver plated brass. 2" x 1.5". $175-225.

FLAMIDOR, ca. 1939. French made chrome plated brass lift arm lighter marked "souvenir of Lourdes" in enamel. Pull out sleeve for re-fueling. $50-75.

FLAMINAIRE, ca. 1948. France. One of the earliest gas lighters. This is a leather covered model. It was non-refillable and used a disposable gas cartridge. Produced by Quercia. $40-60.

HERMES, ca. 1932. France. An 18kt white gold, side-lifting cap, petrol lighter with a top wheel. 1.8" x .7". $1,200-1,500.

FLAMINAIRE, ca. 1948. France. Earliest gas lighter. chrome plated brass with a disposable gas cartridge. To light, one must hold down top button while turning thumb wheel. $30-50.

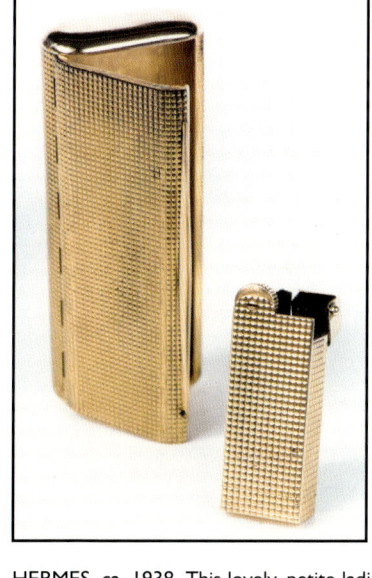

FLAMINAIRE, ca. 1948. France. One of the first gas lighters, this particular model is finished in black lacquer. $40-60.

HERMES, ca. 1938. This lovely, petite ladies set is beautifully designed and implemented. It is 18kt gold. The lighter is marked "HERMES PARIS" on the bottom. It is hallmarked under the cap which is the typical location on French lighters. The case does not bear a name, but is properly hallmarked. Although retailed by Hermes, this set most certainly came from Van Cleef & Arpels. The lighter's design in unmistakable: the sideways lift cap, the pattern of the gold (almost hobnailed), and the incredible hinged bottom for access to both flint and fuel. Invisible to the naked eye the is a hinge six mm up from the bottom of the lighter. This enables the bottom to pop open. Even this hinged bottom is hallmarked for 18kt. $3,500-4,000.

HELIOS, ca. 1935. A table lighter similar to the Allverne lighter. However this one was surely more affordable. It is constructed of plastic and base metal. It bears French tax stamps and is marked "Helios" and "France and Etranger (abroad)." $1,300-1,700.

HERMES, ca. 1942. sterling silver engine turned petrol lighter with a removable fuel tank. The front has an attached chain with anchor in gold, decorated with three sapphires. 2.4" x 1". $850-1,000.

HERMES, ca. 1950. Two gold plated brass petrol lighters. The one on the left is semi-automatic and opens upon a turn of the thumb wheel. The one on the right is a manual version. *Left:* $500-1,000; *right*: $250-350.

LANCEL, ca. 1931. A chrome plated brass "Automatique" pipe lighter. It has an unusual system in that the button on the side can be pushed in like an air pump. This creates a "torch-like" flame similar to the Beattie Jet lighter. 2" x 1.7". $750-850.

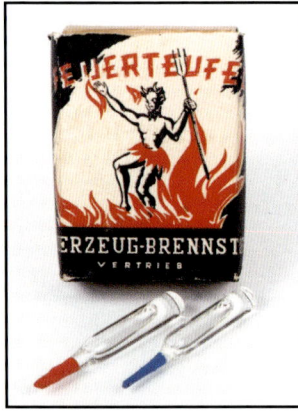

HOFFMAN, ca. 1900-1910. An interesting packet of twelve glass tubes (six of each color) filled with lighter fluid. These are of German manufacture as evident from the excellent workmanship. "Feuerterffel" translates to "Fire Devil" and "Feuraeug-Brennstoff" means "lighter refill" $500-750.

LANCEL, ca. 1935. A semi-automatic lighter with a black and red enamel design on silver plated metal. $600-700.

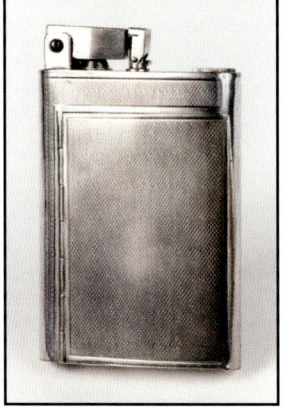

LANCEL, ca. 1936. A silver plated cigarette case with a semi-automatic lighter. This model fills from the top unlike other Lancels. 3.9" x 2.4". $400-500.

LANCEL, ca. 1930. Two Lancel table lighters with identical well recognizable mechanisms, one lighter in copper the other in chrome. Although they appear to be the same, there are a number of differences. Both have eight-day clocks, although the copper version has a dual time zone feature and a day-of-the-week pointer and is an alarm clock as well. The copper one also refuels through a completely different type of screw (it is small and not visible unless the lighter is turned upside down.) In contrast, you can see the large fuel screw at the bottom right of the chrome version. Similarly you cannot see the clock winder on the copper version but you can at the bottom center of the chrome piece. Because the clock in the chrome version has a simple clock, it only requires a winder which doubles (when pulled out) as the time setting device. The copper model opens from the rear to reveal the various controls for its multi function clock. The clock face on the copper model is marked "LANCEL PARIS." The clock face on the chrome model is marked "8 jours" and in tiny letters at the bottom of the dial, "FAB. SUISSE." The only other marking on the chrome version is on the fuel screw: "LANCEL AUTOMATIQUE." On the bottom of the copper version we find "LANCEL" and then in tiny letters "BtdSGDG.75.22" which is the acronym for "Brevete Sans Garantie Du Gouvernement" or "Patent without guarantee from the Government." Also marked "FAB. FRANCE." The copper version with the elaborate clock is quite rare. $1,250-1,500 each.

LANCEL, ca. 1940. A French made, "What were they thinking" lighter. Silver plated engine turned with very unusual (read difficult) flinting mechanism. It appears to have a windguard when closed, but it lifts up when you open the lighter to ignite. 2.5" x .9". $200-250.

LANCEL, ca. 1942. A pair of "Royale," angular, semi-automatic lighters. The smaller one is silver plated brass and the larger appears to be chrome plated aluminum. 2" x 1.4" and 1.8" x 1.2". $150-200 each.

LE FOLLET, ca. 1944. A pair of unusual pink gold plated lighters that were made in Paris, France. They have a semi-automatic action. *Plain:* $75-100; *bicyclist:* $150-175.

LANCEL, ca. 1942. A pipe lighter in silver plate and made in France. Engine turned design and a swing out wick for the pipe. 2.7" x .9". $200-250.

LOWELL, ca. 1922. France made lighter sometimes seen as "Juvenia" and "Mondial." This early lift arm lighter is unique because of its horizontally mounted flint, and also its lift arm which is set diagonally to the body, rather than parallel. $175-225.

LANCEL, ca. 1942. A pair of French made lighters that resemble Dupont lighters. With the cap closed, one has an exposed flint wheel and the other has a fully covered wheel. 2" x 1.3". $200-250 each

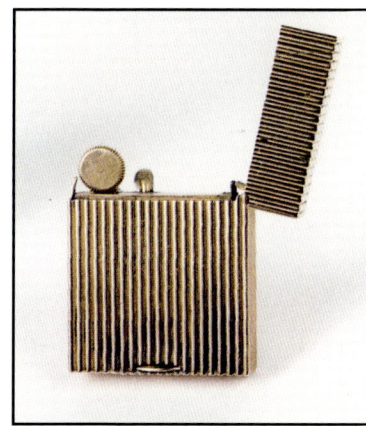

MARCHAK, ca. 1939. A French made 18kt gold single wheel manual lighter with a hinged cap and removable fuel tank. The jeweler's mark indicates that it was made by Marchak. This lighter is similar to an Evans model from that period. 1.7" x 1.4". $1,500-2,000.

LANCEL, ca. 1948. Lancel metal lighter in the shape of a car about 3" long. Also produced in gray. One of the nicest car lighters we've seen. $1,000-1500.

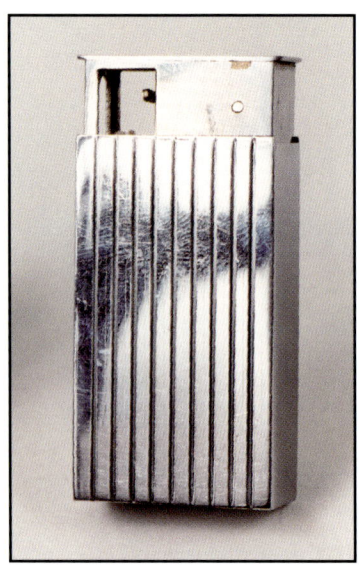

MOKIN, ca. 1934. France. A "Match," chrome plated, matchbox shaped, semi-automatic lighter. 2.3" x 1.4". $1,250-1,500.

PIERRE CARDIN, ca. 1985. France. A pair of cufflinks with a hand holding a lit lighter design. $100-125.

OSTERTAG, ca. 1938. Paris France. A pair of lighters. The one on the left is in gold with black lacquer work and blue sapphires on lid. The other is sterling silver with sapphires on the lid. *Left:* $2,000-2,500; *right:* $1,000-1,500.

SAVENT, ca. 1934. A well made French aluminum lighter with pull-out tank. Very light weight. 2" tall. $50-75.

PAM, ca. 1940. Two examples of this French made lighter. One has black lacquer and brass with side thumb roller while the other has a semi-automatic mechanism with a small thumb wheel. $250-300 each.

TIFFANY/SCHLUMBERGER, ca. 1955. A solid 18kt gold (including the insert) lighter in the shape of a fish. This was a special order item that you are not likely to find in your quest for quality lighters. 4.5" l. $5,000-7,000.

Trench Art, ca. 1915. An incredible trench art piece with an early airplane on one side and a Zeppelin on the other. 2.5" tall. $100-200.

Trench Art, ca. 1917. France. A wonderful trench art lighter in the shape of Kaiser Wilhelm. 3.25" tall. $250-350.

Trench Art, ca. 1917. A French WW I trench lighter in chrome plated brass with a rooster on front (the national bird of France), and Joan of Arc on the reverse side. 2" diameter. $100-150.

Trench Art, ca. 1917. Book shaped trench piece with lighter hidden inside. Brass. $60-90.

Trench Art, ca. 1918. A nicely handmade larger lighter in brass. Probably French. $60-80.

Trench Art, ca. 1917. France (surmised). Three examples of World War I "trench art" lighters. As the troops lived in elaborate trenches for extended periods of time, some of the more clever and talented ones began to fashion all sorts of paraphernalia out of the brass casings of pistol, rifle, and artillery ammunition. Lighter making was a favorite pass time. The most elaborate of the three is the gun. First you manually pivot the rear cover to the open position. This exposes the wick, encased in a machined brass windguard. It functions with a rasp mechanism. When the trigger is pulled, the rasp turns a gear which turns a typical flint wheel which is hidden from view by the front cover. The lighter in the rear of the photo also incorporates a flint and wheel mechanism. But rather than employing a rasp, it ignites when you sharply spin the large knurled knob seen extending from the bottom. A clever lighter, but not well made. Last is "Das Boot." This is a striker lighter. Withdraw the spike from the helmet and it's a typical striker. The ignition material is hidden in the heal of the boot. This is a well made piece. *Left:* $500-750; *center:* $100-150; *right:* $250-300.

Trench Art, ca. 1942. A lighter showing great workmanship made from a military shell and some spare brass. The snuffer arm is shown both up and down. 5.5" across. $250-300.

Unknown, ca. 1850. France (surmised). On the left is a fire steel. It is somewhat crudely made, with an attached flint stone and tinder fusee. On the right is a fire steel. It has a recursive tang and a wedge shaped body of steel, chiseled with small dots and notches on the edge. Smaller end is thicker and incorporates a tool for removing caps. *Left:* 2.6" x .7"; *right:* 3.1" x 2". $400-600 each.

Trench Art. A group of WWI and WWII era trench art lighters. These are French and English and the flint wheel unscrews for refueling. $50-100 each.

Unknown, ca. 1910. Two French made pocket strikers: *Left:* 18k gold, with blue and white enamel. Hallmarked on the edge of the case. The wand bears a large blue sapphire cabochon. The wand is hallmarked and engraved with the word "DÉPOSE." Elaborate engine turning in the gold with enamel accents and an attached ring. *Right:* Another similar 18k gold striker. Hallmarked on the edge of the case. No enamel but even more elaborate engine turning with a similar ring attachment. Monogrammed "JHA." The wand is marked "DÉPOSE." *Left:* 2" x 1"; *right:* is 2.1" x 1". $750-1,250 each.

Unknown, ca. 1800 & 1820. On the left is a combination fire steel and tool. The grip is thick and coarsely cross cut for better handling. May be French, though many of these combination sets are from Spain. It includes ember tongs and a swivel out pipe tool. Well designed with a tension spring to clamp down firmly on an ember. On the right is an interesting pocket fire steel and case. It has a rudimentary push button release and hinged flap which contains a compartment for flint stone and fodder. Engraved, as a flint lock gun might be engraved, with a bird dog on the front and a flower on the back. Both the flap and the main piece have an identical, crudely engraved mark on the inside that could be "VII". *Left:* fire steel with tool measures 3.2" (with tool extended) x 4.3" wide. $1,000-$1,500; *right:* pocket fire steel and case 2" x 1.5". $400-$600.

Unknown, ca. 1913. This sterling silver striker model was made in France. $150-200.

a. 1916. A French blue translucent enamel over engine turned sterling silver, striker lighter with a watch. The lighter shows the remnants of a gold wash. 2" x 1.3". $3,000-3,500.

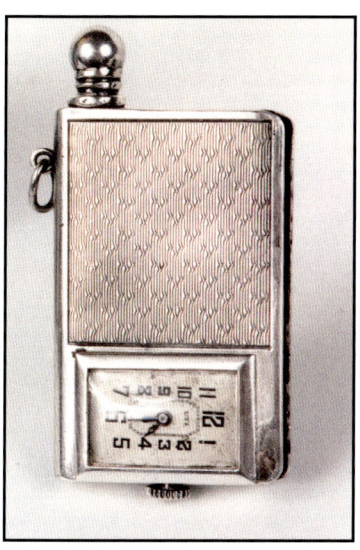
Unknown, ca. 1916. A French pink enamel over engine turned sterling silver, striker lighter with a watch. The lighter shows the remnants of a gold wash. 2.2" x 1.1". $2,500-3,000.

Unknown, ca. 1919. A very fine French made silver and enamel combination striker lighter with fusee. It has an American flag on the front and a French flag on the reverse side. $500-700.

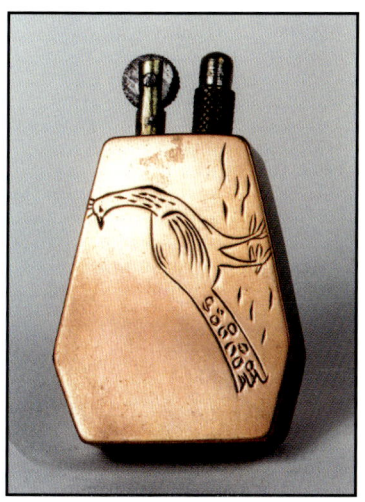
Unknown, ca. 1917. Believed to be French. A World War I era copper and brass trench lighter with an engraved peacock. $125-175.

Unknown, ca. 1920. A striker lighter made in France and marked "Allameur Falque." Interesting ink-well shape. $200-300.

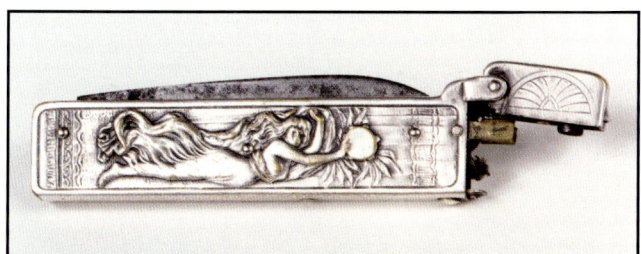
Unknown, ca. 1918. An unusual pocket knife with two blades, semi-automatic flint lighter with European mechanism. Silver plated over brass with elaborate relief scenes — flowers on one side and a virgin holding what appears to be a mirror overhead — on the other side. There are no markings of any sort. 3.7" x .9". $1,500-1,750.

Unknown, ca. 1922. France. Mother of pearl and brass lighter. 1" h. 1.5" w. $60-90.

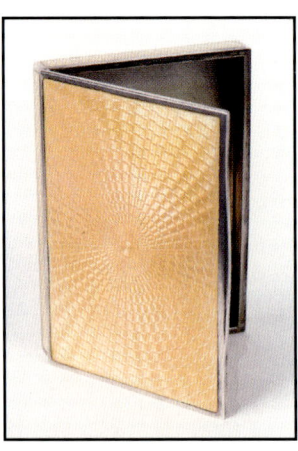
Unknown, ca. 1925. A sterling silver cigarette case with a beautiful yellow enamel face. Made in France. It is 2.6" x 3.8". $500-750.

Unknown, ca. 1933. This small ladies lighter is French, bearing all of the correct hallmarks for both gold (18k) and platinum (the diamond shaped ornamentation running down the center. You simply lift the cap and spin the wheel. It is marked in French as anglicized: "Patent without guarantee from the Government." It also bears several work numbers, and the words: "Mellerio dits Meller."

Unknown, ca. 1928. France (surmised) A .900 silver combination lighter and cigarette case. 3.8" x 2.8".

Unknown, ca. 1933. France (surmised). An Art Deco lift arm table lighter. Chrome plated with black inlaid enamel. It bears no markings other than the French import tax stamp on the bottom of the lighter. 2.8" x 2.2". $200-300.

Unknown, ca. 1930. Probably of French origin, a lift-arm lighter with a large "Lynda" watch and an unusual enamel finish. Chrome-plated brass. Bears a French lighter tax stamp on the snuffer. 2.25" tall. $200-250.

Unknown, ca. 1933. A silver cigarette case with a decorated white enamel face. Made in France. 2.4" x 3.5". $500-750.

Unknown, ca. 1930. An unusual French made, urn shaped, striker lighter on a marble base. The diameter is 5.5". $300-450.

Unknown, ca. 1935. France (surmised). A lady's compendium. It is well made and looks as if it were made for a movie star. Engine turned sterling silver on four sides, both ends are 14kt gold. Ruby and gold finger releases are on both the lighter and the lipstick. The center portion on top contains a compact. The lower section contains a cigarette case. 2.1" x 3.2". $3,000-5,000.

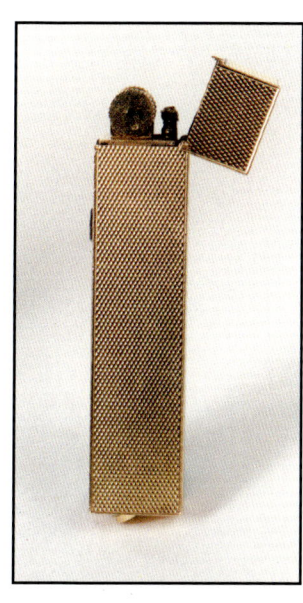

Unknown, ca. 1936. Believed to be French made. 18kt gold single wheel manual lighter with a hinged cap. 2.8" x .6". $750-1,000.

Unknown, ca. 1935. France. This gold plated lighter uses a large thumb roller. $50-75.

Unknown, ca. 1935. France. Sterling silver and gold Dunhill type lift arm lighter. 2" x 1.4". $750-1,000.

Unknown, ca. 1937. This is an 18k gold French lift arm lighter, hallmarked on the cap, the base and the fuel screw. The lighter is beveled and elaborately engraved on all six sides. The flint and wick line up with the front bevel rather than the front of the lighter. To re-flint you pull the flint wheel mechanism out. There is a gold leaf spring which provides gentle tension. The leaf spring fits into a tiny groove within the lighter housing so as to assure that when re-inserted, the flint wheel is properly aligned. $2,000-2,500.

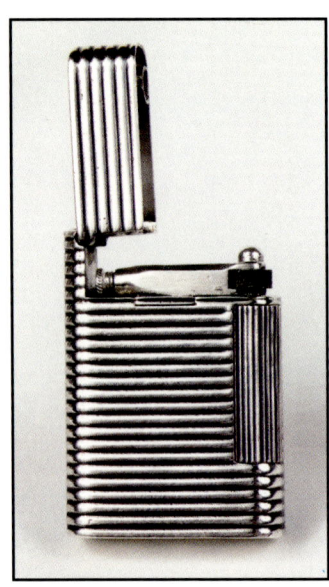

Unknown, ca. 1948. Possible Van Cleef & Arpels. sterling silver lighter with a manual roller bar. 1.9" x 1.5". $500-700.

VAN CLEEF & ARPELS, ca. 1939. France. An exquisite matched heavily ribbed case and lighter by Van Cleef & Arpels. In excellent condition. Both pieces hallmarked for .800 silver. The lighter also bears a work number 55971. The case opens with a perfectly concealed button. The hinges are also well concealed. $3,000-3,500.

VAN CLEEF & ARPELS, ca. 1926. France. The "Paradoxe" model, this is an 18k gold lighter with a built-in match box. If the lighter failed to light, it could be opened and a match removed to light one's cigarette. 2.3" x 1.6". $17,500-$22,500.

VAUDAINE, ca. 1890. A pair of "Spinners." *Left:* sterling with a gold wash. Elaborate engraving and hallmarked on the inside of the cap and further marked "GATON BTF S.D.S.G. PARIS;" *right:* Appears to be nickel plated brass. This is the so-called iron/crystal version (the wheel is iron and the striking material is a flint rock crystal). A complicated set of gears gets the mechanism spinning and a spark will ignite the tinder. 2" in diameter.

VAN CLEEF & ARPELS, ca. 1939. France. An 18kt gold oval lift cap lighter with a long roller bar. The flint tube can be swung away from the body allowing access to the fuel compartment located beneath it. 1.9" x 1.3". $1,200-1,500.

VULC AUTO, ca. 1934. France. A silver plated brass, semi-automatic lighter. 1.2" x 1.6". $200-300.

Great Britain

ASPREY, ca. 1926. Three lighters, two 9kt gold Asprey Wafer lighters, and the thrid (in the center) stamped "Dunhill," made in England. The one on the left is the standard size. *Left:* $600-800; *center and right:* $750-1,000 each.

ALLVERNE, ca. 1933. sterling silver table lighter with battery powered electric ignition for wick and fluid. Hallmarked London, 1933 and bearing the maker's mark "DBH." The lighter bears an engraved name believed to be the name of this model of lighter "ALLVERNE" in quotation marks. It also bears a patent number and a registration number. $1,500-2,000.

ASPREY, ca. 1929. sterling silver table lighter from Asprey. Beautifully engine turned to look just like a small book when closed. Even the spine shines, as it repeats the motif of nicely bound leather books. You slide the spine as shown to activate a typical wheel and flint mechanism. Repeatedly marked "Asprey London" and "A Co. Ltd." and hallmarked for London 1929. $2,500-3,000.

ALVEM, ca. 1930. English made Lighthouse built entirely of pewter. Operated with one dry cell battery and lighter fluid. A battery provided the charge necessary to heat the coil which in turn ignited its petrol soaked wick. 6". $250-350.

ASPREY, ca. 1938. A lovely sterling silver concealed lighter with a slide out mechanism. Large size. $1,400-1,800.

ASPREY, ca. 1916. England. A sterling silver automatic with an elaborate mauve glass enamel over engine turning. Most probably manufactured by RK (Richard Kohn). 2.3" x 1.4". $1,500-2,000.

ASPREY, ca. 1938. Asprey is an English manufacturer specializing in very high quality goods, often in precious metals. This is a concealed lighter which slides out of its case. The one on the left is 18k gold and the one on the right is sterling silver. *Left:* $2,500-3000; *right:* $1,400-1,800.

ASPREY, ca. 1939. England. 9kt gold single wheel, concealed mechanism lighter. 1.9" x 1.2". $2,500-3,500.

BENEY, ca. 1927. Two versions of the rather unique Beney lighters. On both lighters you can lift part of the cap just on its own to expose the flint wheel. Then turn the wheel to light the wick in windproof mode. Or you can lift the entire cap (see hinge on left side) to fully expose the mechanism and ignite without the complication of the windguard, and facilitate its use as a pipe lighter. On the left, we see a sterling version with shagreen cover. Both lift caps, the bottom, the fuel screw, and the flint screw are all hallmarked for silver. Inside the large lift cap we find the mark for 1927. And inside the small lift cap we find the mark for London. I suspect there is a full set of hallmarks on the side, but they have been hidden by the shagreen, which might support a notion that the shagreen was an after thought. The bottom of the lighter is also marked: "THE BENEY LIGHTER PATENTED MADE IN ENGLAND." The larger one (on the right) is also sterling, but it is quite unusual. It was made in London and exported to Paris where it was sold by Hermes. The hallmarks tell the story. Both caps, the side of the lighter, the bottom, the fuel screw, and the flint screw are all hallmarked for silver. The date stamp (this one is 1926) is found on the side and under the large cap. The mark for London is on the side and under the small cap. This one also bears a maker's mark "RB" on the top of the mechanism and the side of the case. Perhaps its large size warranted an outside maker to fabricate the lighter. There are also French tax stamps (rather than the bulky type of stamp affixed to common lighters, this one bears the small rectangular "tax paid" imprint). One is visible on the lower right side of the cap and two on the top of the lighter. These marks indicate not only tax paid but also sterling silver. The bottom of the lighter is marked "THE BENEY LIGHTER MADE IN ENG" followed by "BCM Z428 (possibly a French marking) AND PAT No253570." The fuel screw is marked HERMES PARIS." $500-750 each.

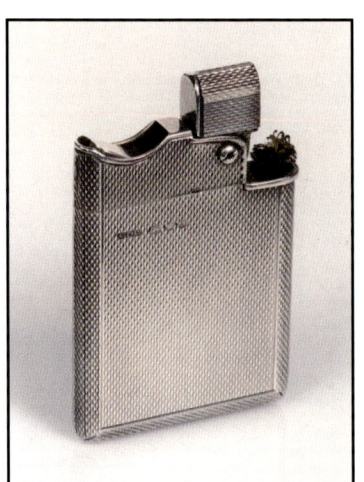

ASPREY, ca. 1946. Asprey is an English maker of high quality clothing and accessories. A sterling silver "Wafer" lighter. 2.175". $375-475.

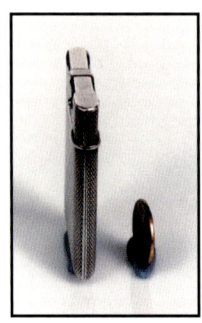

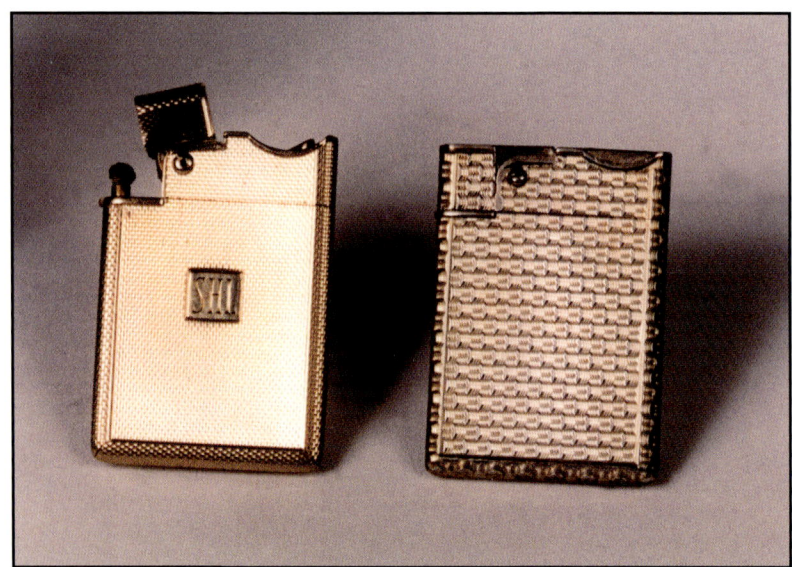

ASPREY, ca. 1948. The model on the left is 9kt gold, while the one on the right is sterling silver. *Left:* $1200-1600; *right:* $375-475.

BENEY, ca. 1928. 9kt gold version of this well designed lighter. Being crooked, the initial plate is probably a later "add on." The center section of the cap can be lifted to expose the flint wheel (which is hidden behind the bulge facing you in the picture. Then rotate the wheel which ignites the wick that is protected from the elements by the engine turned cap. On a nice day, just lift the entire cap and ignite the exposed wick. Nicely engine turned. When the fuel screw is removed you see attached a very sharp needle used to pull up the wick. Marked on the cap, bottom, and on the fuel screw ".375." Also on the bottom "THE BENEY LIGHTER, PAT NO 253570, MADE IN ENGLAND." On its side, the lighter is hallmarked four times: R. B. (maker's mark), .375, 1928-29, London. $3,500-4,500.

CALVERLEY, ca. 1930. An English made, silver, slide action lighter. 2.3" x 1.4". $400-600.

CHARLES, ca. 1950. "The Charles" lighter was made in England. It is silver plated and is a beautifully engineered lighter. The photo shows its original box which includes instructions and tool for adjusting the wick in the "castellated" wick nozzle. The small lift cap houses a three-barrel holder that holds the main flint and two spares. A spare is loaded simply by opening the cover to the barrel housing the flint currently in use, removing the spring and any remnant of the flint, rotating the barrel mechanism (as in a pistol revolver), inserting the spring mechanism and closing the cover. Comes in the original box with all the fittings. 1.9" x 1.7". $400-500.

CLASSIC, ca. 1927. English made lighter in sterling silver with unusual side mounted thumb and flint wheels. $300-350.

CHARLES, ca. 1949. England. A beautifully designed lighter produced by Charles Engineering Co. The vertical thumb roller creates the spark while also rotating the flint chamber for even wear. 2" h. $400-500.

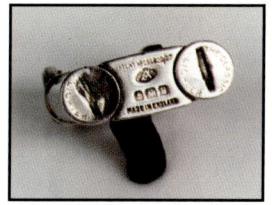

CLASSIC, ca. 1928. An English made sterling silver "ball shaped" lighter. It has an unusual vertically mounted thumb and flint wheel. 4" h. $600-800.

COLIBRI, ca. 1929. A pair of very rare "Original" models with a hidden slide to hold a photo. $1,250-1,500 each.

CLASSIC, ca. 1932. Table lighter, made in England of nickel plated brass. Uncommon model and most often seen in a cylindrical shape. $200-300.

CLASSIC, c.1935. England. A British-made lift-arm lighter with an unusual vertical spark wheel. Made of sterling silver with a polished shagreen wrap. 1.8". $350-450.

CLASSIC, ca. 1939. A handsome man's pocket lighter made of sterling silver and marked with the "County of Warwick Bombing Squad." Made in England. $300-400.

COLIBRI, ca. 1929. First Colibri model. A chrome "Original" sometimes referred to as the "Kick-start" lighter with a celluloid wrap of the city of Koln. "Colibri" is the name of a Central American bird. $150-200.

  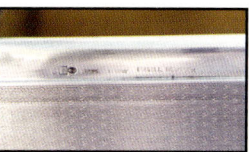

DUNHILL, ca. 1924. Exquisite and rare sterling silver Dunhill "Unique" lighter and cigarette box. Produced in Germany for Dunhill, New York. Only one of its kind known at this time. Lighter slides off for easy re-fueling. Inside lined with cedar. Measures 7" across and 4" deep. $7,500+ up.

COLIBRI, ca. 1942. The "Famous" model. This semi-automatic lighter is activated by turning the wheel which releases the snuffer cap. 2" tall. $125-150.

DUNHILL, ca. 1924. A 14kt gold "Unique B" model with an engine turned design. $1,000-1,400.

DUNHILL, ca. 1919. England. A "Column" model. Very early Dunhill table lighter made of nickel plated brass with black Galbraith (a bakelite-type of material) covering. This model was also produced by Dunhill in a pocket version. $1,000-1,400.

DUNHILL, ca. 1924 England. A sterling silver (marked W & G) lift arm, "Unique B" lighter. The design is a cinnabar red enamel with a Chinese design. 2.1" x 1.7". $4,500-7,500.

133

DUNHILL, ca. 1925. England. A sterling silver (marked W & G) lift arm, "Unique B" lighter. The design is a hard glass pale blue enamel over an impressed design. 2.1" x 1.7". $4,500-7,500.

DUNHILL, ca. 1925. England. A Dunhill set – a lighter and cigarette case – in sterling silver with a green and black enamel decorated with marcasites. The lighter is a "Unique A" marked "W&G." The lighter is 1.8" x 1.5". $7,000-10,000.

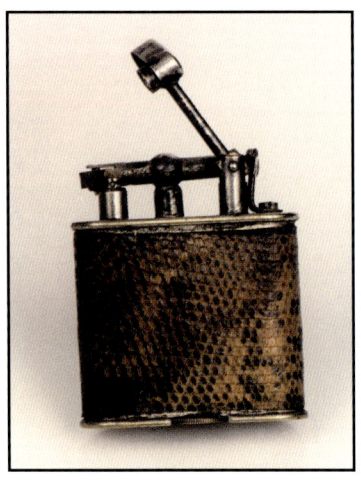
DUNHILL, ca. 1925. A "Unique A" model covered with snakeskin. $100-150.

DUNHILL, ca. 1926. A sterling silver lift arm watch lighter with an engine turned design in its original box. 1.8" x 1.5". $3,000-4,000.

DUNHILL, ca. 1925. A pair of 9kt gold lift arm watch lighters. These are exceptionally rare self winding models that are automatically wound each time the lift arm is raised. *Left:* 1.8" x 1.5"; *right:* 2.1" x 1.7". $10,000-15,000 each.

DUNHILL, ca. 1926. Pages from the Dunhill 1926 catalog showing decorative lighters and cigarette cases.

134

DUNHILL, ca. 1926. A sterling silver lift arm, "B" size lighter. The design is a hard glass light green enamel over engine turning in a sunburst design. 2.1" x 1.7". $4,500-7,500.

DUNHILL, ca. 1926. A sterling silver, "Unique B" size, lift arm lighter with two color (yellow and pink) 14kt gold stripes. $400-500.

DUNHILL, ca. 1926. A Dunhill one piece cigarette lighter and cigarette case in sterling silver with engine-turned decoration. 4" x 2.3". $7,500-10,000.

DUNHILL, ca. 1927. England. Unique lighter and cigarette case in sterling silver with black enamel and marcasites. $2,500-3,000.

DUNHILL, ca. 1926. England. A Dunhill 14kt gold cigarette case with a concealed "Unique" lighter. 2.3" x 3.5". $8,000-10,000.

DUNHILL, ca. 1927. An sterling silver lift arm, "A" size lighter. It has blue enamel work and an attached RAF badge. 1.8" x 1.5". $1,500-2,000.

DUNHILL, ca. 1926. England. A Dunhill three piece lighter grouping including a "Unique B" lighter with a cigarette holder and cigarette case. Beautiful sterling silver and enamel. Lighter, $4,500-8,000; cigarette holder and case, $2,000-2,500.

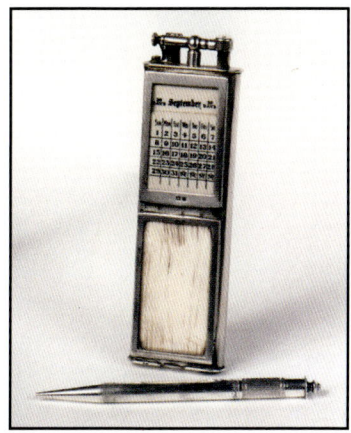
DUNHILL, ca. 1928. A very unusual, "Bridge Lighter," sterling silver, club size, lift arm lighter. It doubles as a perpetual calendar and bridge scorer with incorporated mechanical pencil (which slides into a hole in the bottom). The calendar and writing surface are ivory and the pencil is sterling silver. Marked "Cartier" and retailed by them. 5.8" x 1.8". $7,500-10,000.

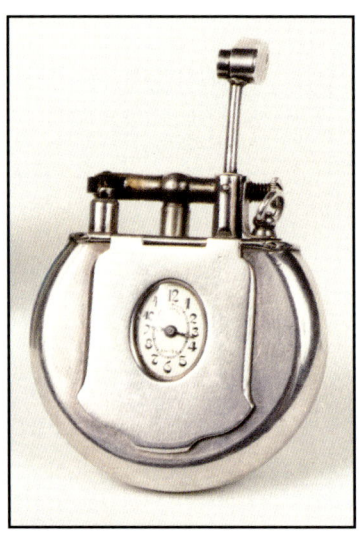

DUNHILL, ca. 1928. Switzerland. An extremely rare sterling silver circular Dunhill watch lighter "Fancy Shape." Oval watch opening. 2" x 1.8". $10,000-15,000.

DUNHILL, ca. 1928. Switzerland. Four sterling silver lift arm watch lighters with different engine turned designs. All "Unique B" models. 2.1" x 1.7". $2,500-4,000 each.

DUNHILL, ca. 1928. Switzerland. 1927. A sterling silver, "Fancy Shape," two tone blue enameled, lift arm, watch lighter in an unusual three quarter oval shape with corresponding oval watch. 2" x 1.8". $20,000-25,000.

DUNHILL, ca. 1928 and 1925. Two lift arm, "Unique A" lighters. The one on the right is done in sterling silver with a green bird (marked W & G sterling), the one on the left has an oriental motif and is Swiss made in silver plate. 1.8" x 1.5". $3,500-5,000 each.

DUNHILL, ca. 1928. A 14kt gold lift arm, "A" size lighter. It has a single wheel and wonderful two tone (pink and yellow) gold design. 1.8" x 1.5". $1,500-2,000.

DUNHILL, ca. 1929. England. "Unique Curve" lighter. This rare, slightly curved model was designed for a vest pocket and made to resemble a small curved body flask. It is silver plated brass. 2" tall. $400-600.

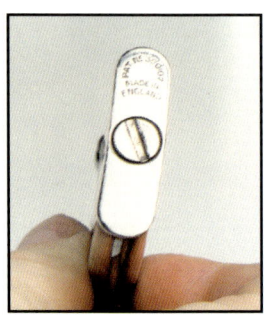

136

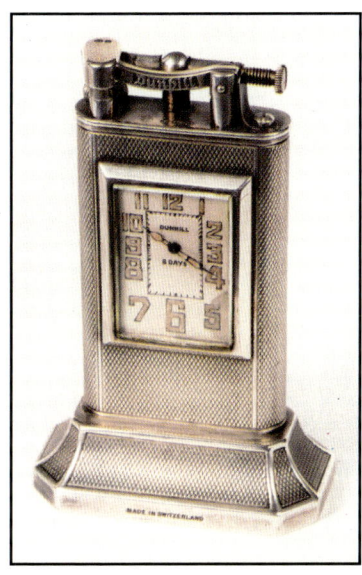

DUNHILL, ca. 1929. Switzerland. A combination table lighter and eight-day clock. sterling silver with engine turning. 3.9" x 2.4". $8,000-10,000.

DUNHILL, ca. 1929. Switzerland. Two lift arm, "Unique A", Art Deco, sterling silver lighters in black enamel with gold accents. 1.8" x 1.5". $1,000-2,500 each.

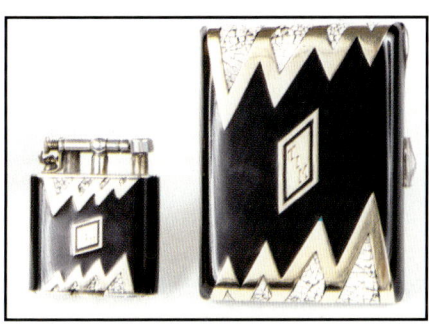

DUNHILL, ca. 1929. Switzerland. A rare "Unique A" lift arm lighter set with a watch in black lacquer and eggshell over sterling silver with a gold inlay (shown front and back) with a ring to attach it to a chain. If the lighter was not incredible enough, it has a matching cigarette case. This is the only known Dunhill enamel set with watch at this time. Lighter, 1.8" x 1.5", case, 3.1" x 2.2". $25,000 and up.

DUNHILL, ca. 1929. Switzerland. A lift arm, "Unique A", Art Deco lighter in sterling silver. The map of the world is, at first, difficult to appreciate. On further examination, it is "Eurocentric". 1.8" x 1.5". $2,000-2,500.

DUNHILL, ca. 1929. Switzerland. A combination table lighter and eight day clock in plain sterling silver with a gold monogram. 3.9" x 2.4". $8,000-10,000.

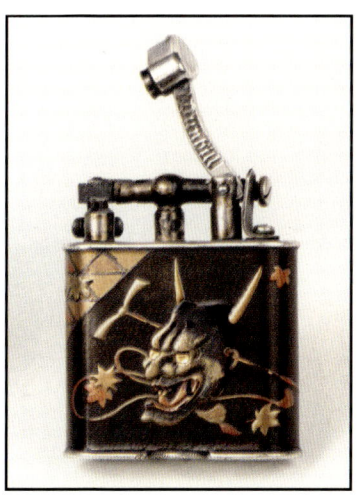

DUNHILL, ca. 1929. Switzerland. A sterling silver Dunhill Namiki lift arm, "Unique A" lighter. It has a very rare devil image. Value undetermined. Identical image front and back on a Dunhill fountain pen sold at auction before add-ons for $100,000. 1.8" x 1.5".

DUNHILL, ca. 1929. Switzerland. A 9kt gold beveled lift arm watch lighter. 1.8" x 1.7". $7,500-8,500.

DUNHILL, ca. 1929. A sterling silver "Unique A" model with a built-in watch. $2,000-2,500.

DUNHILL, ca. 1929. A sterling silver lift arm watch lighter in black and blue enamel. It is a unique fancy shape, covered with hard glass enamel and an elaborately engraved St. Christopher design on the back side. It is marked on the base "Dunhill London. Swiss made." 1.9" x 1.7". $25,000-30,000.

DUNHILL, ca. 1929. A "Sports Bijou" model in 9kt gold. $1,500-2,000.

DUNHILL, ca. 1929. Sterling silver with powder blue, translucent, hard glass enamel work, lift arm watch lighter. 1.8" x 1.5". $5,000-8,000.

DUNHILL, ca. 1929. A pair of "Unique B" models, both in 14kt gold with a watch. On the left is a pretty, engine turned example with a horizontal watch. It is a single wheel model with straight snuffer arm. The model shown on the right has a double thumb wheel and a curved snuffer arm. It has a vertical watch with a plain finish. *Left:* $3,000-3,500; *right:* $2,500-3,300.

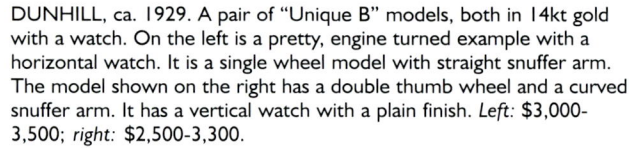
DUNHILL, ca. 1929. A 9kt gold, octagonal, beveled lift arm watch lighter with sapphire decoration. It has an unusual detail to open the flap which accesses the watch area. 1.8" x 1.5". $7,500-10,000.

DUNHILL, ca. 1929. Stunning 14kt large "Sports" model with watch. It is marked 14c on the top which, along with the bottom marking of its US patent number, indicates it was made for retail in the US It is accessible by using the thumb piece at the top to drop the watch panel down. Back of watch case marked "PATENT APPLIED FOR," "SWITZERLAND," "DUNHILL," and "14c." It bears the serial number "590" as does the front of the watch case (hidden by the flap). The flap and the case, on the other hand, bear the number "581." Usually all four numbers match. Possibly two lighters were cobbled together to create this one. The bottom is marked with the British patent number, "MADE IN SWITZERLAND," and "OTHER PATENTS PENDING." Beautifully engine turned. $6,000-7,500.

DUNHILL, ca. 1929. A 14kt pink gold, lift arm, "Long Bijou" lighter. This was probably detailed by Cartier or other jeweler since the lift arm is not marked. 1.3" x 1.9". $1,000-1,300.

DUNHILL, ca. 1930. Beautifully enameled "Unique A" model made in Paris. sterling silver with curved snuffer arm. Thumb wheel not original. $3,500-5,000 (with correct thumbwheel).

DUNHILL, ca. 1930. "Unique A" size in sterling silver and enamel with an oriental design. $3,500-4,500.

DUNHILL, ca. 1930. A Dunhill "Compendium" in sterling silver consisting of a lighter and a cigarette case with a watch. The case is decorated with black enamel and marcasites. Made in Paris. 2.7" x 4". $4,500-7,500.

DUNHILL, ca. 1931. England. A 9kt gold lift arm watch lighter and its original red box. 1.8" x 1.5". $3,000-4,000.

DUNHILL, ca. 1931. A 9kt gold, beveled, concealed, octagonal watch, lift arm lighter with lapis and black enamel decoration. It has an simple slide locking action to open and close the watch. 1.9" x 1.6". $25,000-30,000.

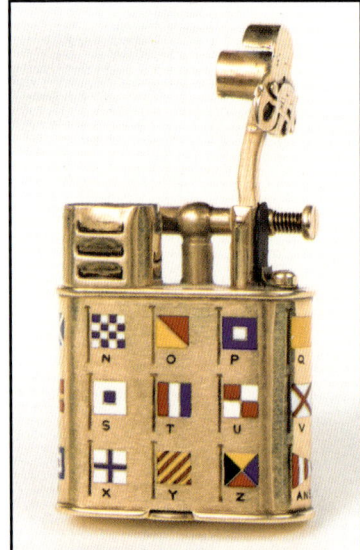

DUNHILL, ca. 1931. One of the most striking 14kt gold with enamel work, lift arm, "Unique B" lighters we have seen. The entire sailing flag alphabet covers all four sides of the lighter and the snuffer has an attached gold initial. This particular model is also marked "Cartier". 2" x 1.6". $5,000-7,500.

DUNHILL, ca. 1931. A "Unique A" model with a curved arm and double wheel. It is made of 9kt gold and sapphires with a watch that sports an unusual eight sided watch bezel. The front opens from the top down for access to watch. $7,500-10,000.

DUNHILL, ca. 1932. Solid 18kt gold "Bijou" lift arm lighter with its original box. 1.3" x 1.5". $2,500-3,000.

DUNHILL, ca. 1931. An 18kt gold lift arm, "Unique A" lighter. It was made in Paris and sports a double wheel. 1.8" x 1.5". $1,500-2,000.

DUNHILL, ca. 1932. France. Two lift arm, Art Deco, French made "Bijou" lighters. One in sterling silver, the other in silver plate, both with black enamel designs. 1.3" x 1.7". *Left:* $500-750; *right:* $1,500-2,000.

DUNHILL, ca. 1932. Dunhill sterling and enamel cigarette case and matching lighter in original fitted box. The Art Deco design features alternating bars in vermeil (gold and silver). The design goes all the way around the lighter and one side of the case; the reverse is black. The lighter is hallmarked with "A.D." in diamond and "London 1930," with silver marks on fuel screw and snuffer cap. The curved snuffer arm has "Dunhill" on it. The case is marked with "G. F." (makers mark) with London sterling silver import hallmarks for 1930 The interior has a gold wash and an inscription for 1932. All is housed in a gold embossed box with Alfred Dunhill Ltd. London & The Prince of Wales feathers marked on the ivory silk. Extremely rare to find boxed deco set, featured in Dunhill 1931 catalogue. $6,000-8,000.

DUNHILL, ca. 1933. Small size 9kt curved bar gold lift arm lighter. This lighter carries an elegant fob with tassels. The monogram is surrounded by a circle containing twenty diamonds and nineteen emeralds. This lighter is in almost perfect condition. $3,500-4,000.

DUNHILL, ca. 1932. Silver plated table lighter with elaborate engraved salutation. $350-500.

DUNHILL, ca. 1933. England. A 9kt gold, lift arm, "Sports A" lighter. 1.8" x 1.7". $1,200-1,600.

DUNHILL, ca. 1932. England. A Dunhill set in blue and black lacquer and sterling silver consisting of a "Unique A" lighter with a matching cigarette case. $5,000-6,000.

DUNHILL, ca. 1933. A gold plated, lift arm, "Bijou Sports" lighter. 1.4" x 1.6". $250-300.

DUNHILL, ca. 1933. "Unique" 14k Gold marked (and retailed by) "Cartier." Produced for the US Market and hand engraved "14K Solid Gold Dunhill" on bottom. $800-1,200.

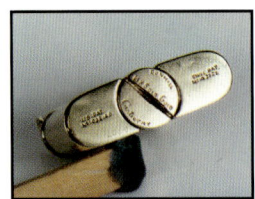

DUNHILL, ca. 1933. A Dunhill cigarette case with a beautiful hard glass enamel showing a pair of peacocks. Marked "Sterling Germany" and "Dunhill, New York". $1,500-2,500.

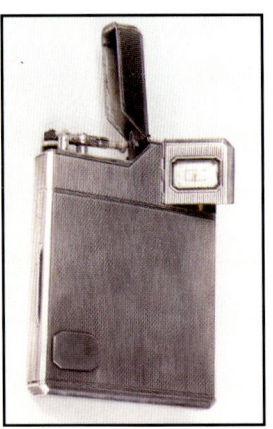

DUNHILL, ca. 1933. France. A Dunhill "Compendium" in sterling silver consisting of a lighter and a cigarette case with a watch. Made in Paris. 3.2" x 4". $3,000-5,000.

DUNHILL, ca. 1934. England. A tall, bevel sided, 9kt gold, lift arm lighter with a sterling silver monogram. 2.4" x 1.5". $2,000-2,500.

DUNHILL, ca. 1934. England. A 9kt gold "Sport" lift arm watch lighter with a windguard. 2.1" x 1.8". $4,000-5,000.

DUNHILL, ca. 1935. A hard-to-find Dunhill "Slim" model upon which the current Dunhills are modeled. Silver plated brass with engine turned design. $175-225.

DUNHILL, ca. 1935. England. "Baby Sylph" Silver plated with engine-turned design. The smallest Dunhill lighter ever produced. 11/16" h. x 7/8" across. $600-800.

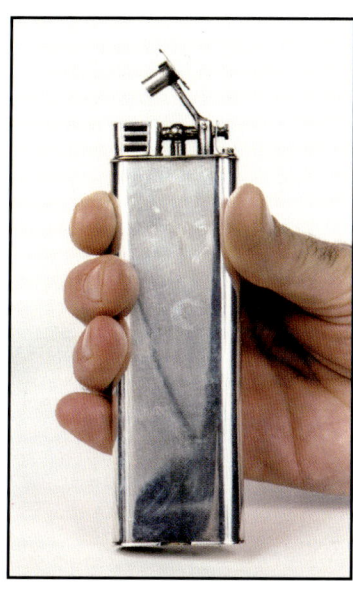
DUNHILL, ca. 1936. A rare "Sports" club sized, silver plated, single wheel, lift arm lighter. 5.8" x 1.8". $1,000-1,500.

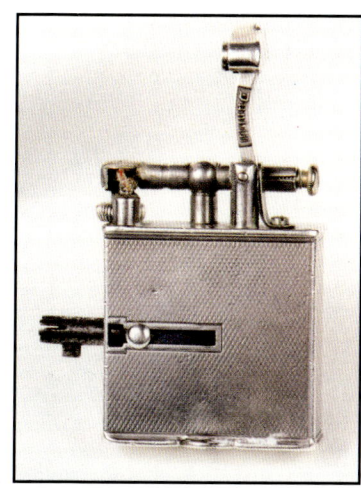
DUNHILL, ca. 1936. Rectangular sterling silver lift arm with a hidden briefcase key and contemporary period brief case. 1.9" x 1.5". $6,000-7,500 lighter only.

DUNHILL, ca. 1936. Switzerland. A sterling silver model with an 8-Day Clock. $6,000-8,000.

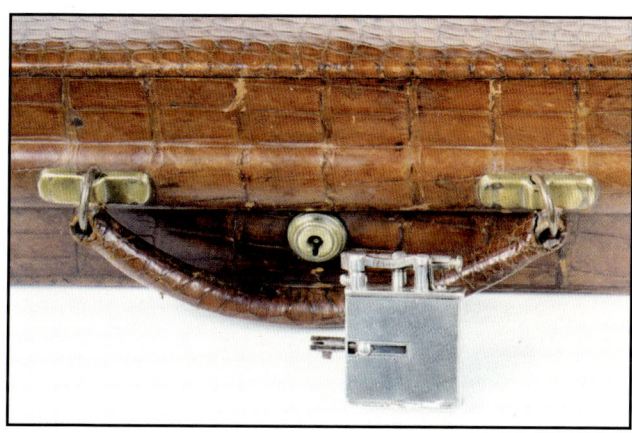

DUNHILL, ca. 1936. England. A combination walnut cigarette box and striker lighter. 5.3" wide x 2.8" tall x 3.7" deep. $600-800.

DUNHILL, ca. 1937. England. A Dunhill "Alduna" 18kt gold lighter. 1.9" x 1.3". $1,000-1,250.

DUNHILL, ca. 1937. 9kt gold "Handy" model with monogram. 1.9" x 1.3". $1,500-2,000.

DUNHILL, ca. 1938. England. A Dunhill Namiki set consisting of a "Savory" lighter and a cigarette holder, both signed by the artist. The lighter is 2" x 1.4".

DUNHILL, ca. 1937. "Ball" lighter. This silver plated ball lighter incorporates Dunhill's double wheel lift arm mechanism. $600-800.

DUNHILL, ca. 1939. The "Baby Sylph," Dunhill's smallest lighter measuring just under one inch. Gold plated engine-turned finish. $600-800.

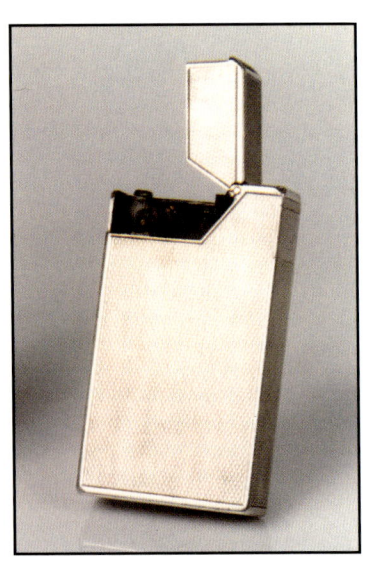

DUNHILL, ca. 1938. The Dunhill "Broadboy" lighter. This is the taller version with a small thumb wheel. It is well hallmarked and was made in England. $400-450.

DUNHILL, ca. 1939. A long silver plated "Wafer Unique" model in an unusual size. This model was slightly thinner than the standard "Unique" model. 2.75" tall. $250-350.

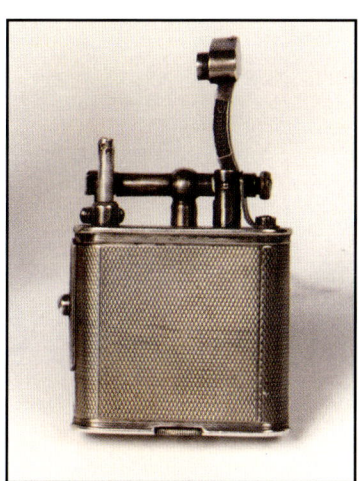
DUNHILL, ca. 1939. Switzerland. A sterling silver Dunhill pipe lighter. The slide on the side will both raise and lower the wick so that it may be used for a pipe. Very rare to find in solid silver; more often seen with plated finishes. $1,200-1,500.

DUNHILL, ca. 1940. An extremely rare table lift arm double wheel silver plated lighter with a blue lizard cover that was made in England. The quarter is shown for size comparison. 6.25" x 3.2". $4,500-5,500

DUNHILL, ca. 1940. USA. A "Silent Flame" lighter. Chrome with a green bakelite wand. It used three "C" cells to function. 4" tall. $150-200.

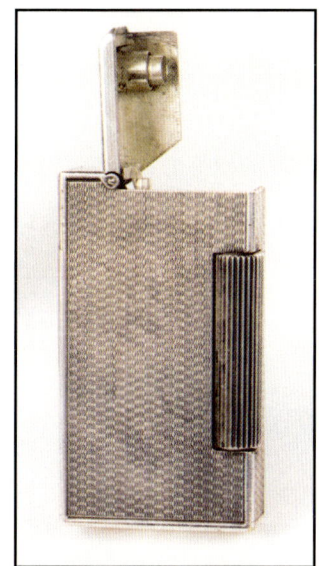
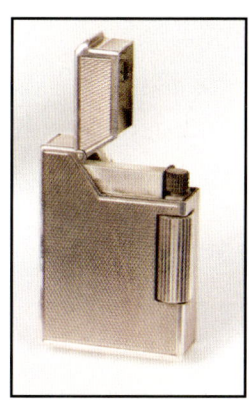
DUNHILL, ca. 1940. A pair of "Broadboy" model silver plated lighters that were made in England. *Left:* 2.3 x 1.3"; *right:* 1.8" x 1.3". $175-250 each.

DUNHILL, ca. 1940. An extremely rare and beautiful "Savory" Club sized (or "Long Handy") Namiki fish design (signed by the artist) lighter with its original box. 3.9" x 1.3". $12,000-15,000.

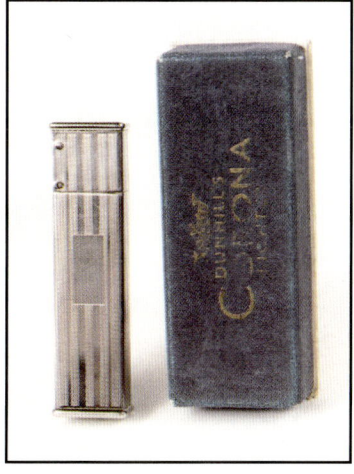
DUNHILL, ca. 1942. A "Corona" silver plated automatic lighter. 2.9" x .7". $250-350.

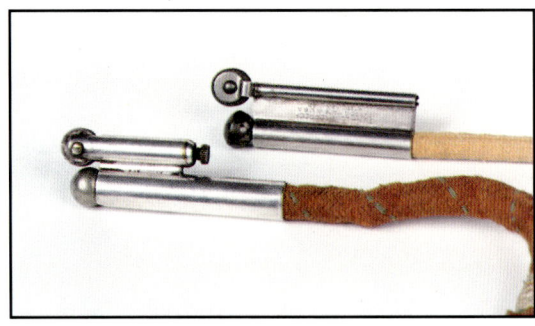

DUNHILL, ca. 1942. A pair of windproof rope lighters, the one shown at the top in sterling silver and the one shown at the bottom in silver plate (marked "Dunhill, London"). *Top:* 2.3" x .8", $400-500; *bottom:* 2.5" by .8". $50-75.

DUNHILL, ca. 1944. England. This is the "Service" model table lighter in silver plate with an engine-turned design. $300-500.

DUNHILL, ca. 1942. A very elegant "Alduna" in sterling silver with 18K rose gold. $400-700.

DUNHILL, ca. 1942. France. "Salaam" black lacquered aluminum model for "Marinha Do Brasil." 2.25" h. $250-300.

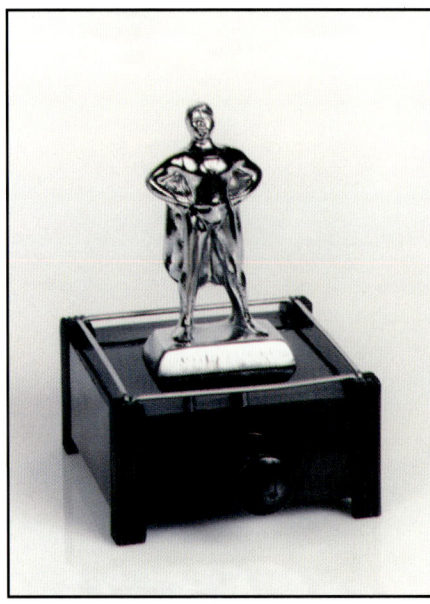

DUNHILL, ca. 1944. USA. An incredible and very rare "Silent Flame" battery operated lighter with a Superman figure. 4.5" h. $1,200-1,500.

DUNHILL, ca. 1943. England. "Handy" silver plated model in its box. The letter "M" is engraved on the snuffer in red. $250-300.

DUNHILL, ca. 1944. "Silent Flame" models used a battery to heat a coiled wire next to the wick when the metal wand was touched to the wire "fence" and the center ornament. This is a replica of the San Francisco Sun Tower. 4" across. $300-400.

DUNHILL, ca. 1946. A pair of "Salaam" model lacquered aluminum lighters. The one on the left is marked Namiki and signed by "Seishou." Both are made in Paris. The roller bar can be reversed because there is a "spare" flint wheel on the opposite end of the bar. 2.3" x 1.3". *Left:* $3,500-4,000; *right:* $250-350.

DUNHILL, ca. 1946. Carltonware made ceramic items. This is a beautiful Carltonware ceramic Dunhill with oriental motif. $1250-1750.

DUNHILL, ca. 1946. A pair of "Broadboy" 9kt gold lighters with the long thumb wheel. 2.2" x 1.3". $800-1,000 each.

DUNHILL, ca. 1947. A "Rollboy" 9kt gold lighter with a diamond shaped monogram in silver. 2.4" x 1". $1,200-1,400.

DUNHILL, ca. 1948. England. A Dunhill "Bell" chrome plated lighter with an ivory-like handle. It is has the same mechanism as the Dunhill "Tankard." 6.5" x 3". $400-600.

DUNHILL, ca. 1948. A pair of Dunhill Aquarium "Service" size table lighters. One is predominantly green on silver plate, one is predominantly blue also on silver plate. The design is reverse painted on a Lucite-like plastic material to give the illusion of water and a three dimensional effect. Word has it that these types of lighters are an early example of recycling. The thick Perspex (type of plastic) was used in the RAF during WW II for windshields. These lighters made good use of the surplus after the war. This is the smaller of two sizes. In both sizes there are four Perspex panels, each of which is reversed carved, hand painted and backed with a type of tin foil to create the reflective nature of the lighter. 2.95" tall x 2.1" wude x 1.4" tall. $2,000-2,500 each.

DUNHILL ca. 1948. A Dunhill "Aviary" table lighter with a bird and forest scene. Aviaries are much scarcer than the more often seen Dunhill Aquariums. Beautifully hand painted in reverse on the insides of each panel. $2,500-3,000.

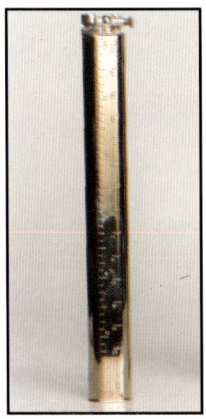

DUNHILL, ca. 1948. A "Sylph Ruler" desk model lighter in gold plate. 6" long. $125-175.

DUNHILL, ca. 1949. A second model "Roman Lamp" model lighter with a similar mechanism to the Dunhill tinder pistol. The finish is a burnished silver plate. 6.5" across. $250-350.

DUNHILL, ca. 1948. England. A "Clipper" gold plated lighter. 2.7" x 1.2". $200-250.

DUNHILL, ca. 1949. Table lighter with clock and covered in alligator skin. The clock is an 8-day version with the dial marked "Dunhill." $4,000-5,000.

DUNHILL, ca. 1949. England. Table lighter with an attractive leather covering. 4" h. $400-500.

DUNHILL, 1950. Page from the 1950 Dunhill export catalog showing the "Sports" and "Pipe" lighters.

DUNHILL, 1950. Page from the 1950 Dunhill export catalog showing the "Tinder Pistol" lighter.

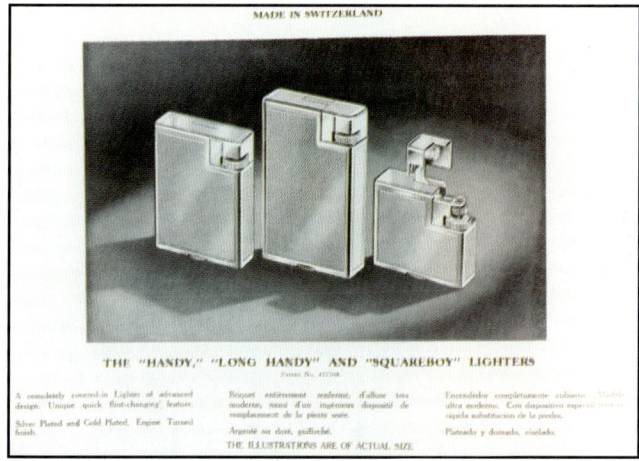

DUNHILL, 1950. Page from the 1950 Dunhill export catalog showing the "Handy," "Long Handy," and "Squareboy" lighters.

DUNHILL, ca. 1952. Another version of the "Pagoda." This one is covered in ostrich and bears a noble family crest – a "C" with a crown on top. $1,500-2,000.

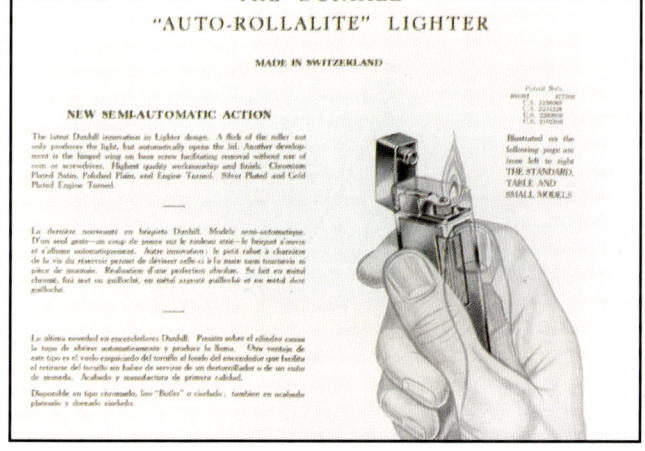

DUNHILL, 1950. Page from the 1950 Dunhill export catalog showing the "Auto-Rollalite" lighter.

DUNHILL, ca. 1952. Still another version of the "Pagoda." This one is ceramic and came in a variety of colors. It stands beside the ceramic shell and was never used. $1,500-2,000.

DUNHILL, ca. 1968. A nice large Brilux table lighter made for and retailed by Dunhill. It was made in Switzerland and used a clever automatic mechanism where the knob on the right side is pressed to release the snuffer arm and cause the spark. It is filled with fluid from the top and is gold plated metal with a leather cover. Also marketed by Guebelin. $500-750.

DUNHILL, ca. 1952. A brass Dunhill table lighter supposedly used by Dunhill as a test form for various Perspex fittings. Also shown is a variety of potential lighter panels. These pieces appear to have come from the estate of one of the Dunhill Museum curators.

DUNHILL, ca. 1953. Switzerland. A pair of "Sylph" letter opener lighters. One is covered in green leather while other is gold plated and engine-turned. 8" l. $275-325 each.

DUNHILL, ca. 1990. England A butane table lighter in gold plated metal. It is similar to the 1920s "Unique" models. 5" tall. $250-300.

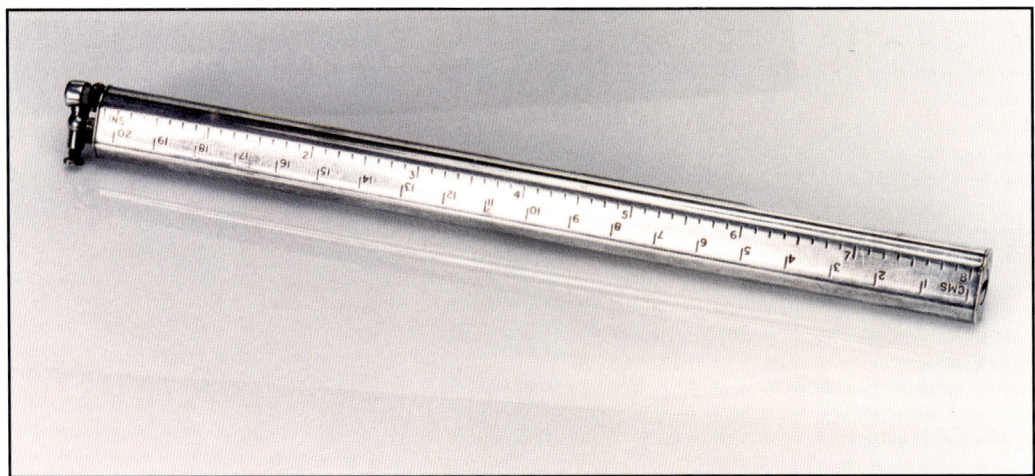

DUNHILL, ca. 1953. Switzerland. A rare size "Sylph" 20 cm ruler lighter. Silver plate over brass. 8" l. $1,200-1,500.

EVEREST, ca. 1933. This Everest lighter was made in England and is sterling silver. 1.675" tall. $125-175.

HAHWAY, ca. 1922. English made semi-automatic Hahway lighter encased in a 9kt gold sleeve. Retailed by Asprey. $400-600.

EVEREST, ca. 1933. Everest lift arm lighters. Made in England. Both have a shagreen finish. $175-225.

JET, ca. 1940. A British-made chrome pipe lighter in an original Jet box. This was the probable inspiration for the Beattie "Jet" Lighter. However it does operate with two tanks. 2.2" x 1.7". $1,000-$1,200.

EVEREST, ca. 1933. Everest lift arm lighters. English made. One is sterling silver and the other is an enamel on sterling silver. *Left:* $250-300; *right:* $400-500.

JET, ca. 1940. A British-made chrome leather covered pipe lighter. Probable inspiration for the American Beattie "Jet" lighter, however it does operate with two tanks. 2.1" x 1.7". $1,000-1,250.

LONCRAINE BROXTON, ca. 1981. England. Very unusual stick of dynamite novelty lighter. The fuse is removed to make contact with spark below for a flame. $20-30.

MANIFOLD ca. 1938. On the left is a most unusual semi-automatic pipe lighter. When ignited it can be used as a windproof cigarette lighter, or by rotating the black wheel (partially exposed on the top left) the wick mechanism rotates up and out allowing for its pipe function. Marked "MANIFOLD," "Pat. No. 135150-13515 Austria," and bears a small triangle with the letters "J. L." Also marked "Alpacca." On the right is another base metal semi-automatic windproof lighter. It is marked "Made in England, Pat. Appl. For." It is also marked "Nura" in script. The fuel tank is extracted from the bottom. There is no fuel screw. $100-200 each.

MCMURDO, ca. 1950. England. An "Ultralite" tall semi-automatic lighter, with a yellow sunburst design in glass enamel, sterling silver sleeve over brass. Note the position of the thumb wheel. McMurdo snuffer caps rise perpendicular to the body rather than parallel as is common with most lift-arm lighters. 2.5" x 1.6". $750-1,000.

MAPPIN & WEBB ca. 1939. Three piece set consisting of lighter, cigarette case, and compact. Beautiful glass enamel on sterling silver. Lighter is German "KW" mechanism. $1,500-2,000.

MCMURDO, ca. 1950. England. A lighter with a shagreen (sharkskin) covering. 2.2" x 1.5". $150-250.

MAURICE, ca. 1933. England. An unusual design for a lift arm lighter. When you handle this lighter you can see that it's mechanism is awkward. That would explain why you don't often see this design. There is a stylized "M" for Maurice on the lift arm cap. The bottom is marked "Maurice London" followed by Provisional Patent and Registration numbers. The lighter is also marked "Foreign." $500-600.

MCMURDO, ca. 1950. England. A short lighter with beautiful glass enamel with sunburst design on a sterling silver sleeve. Brass body beneath sleeve. 1.8" x 1.6". $350-450.

MOSDA, ca. 1933. A sterling silver jacketed, lift arm lighter with marcasites. Made in England. 2" x 1.3". $275-325.

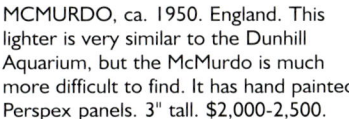
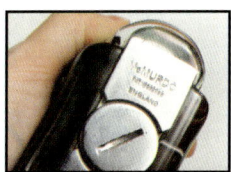

MCMURDO, ca. 1950. England. This lighter is very similar to the Dunhill Aquarium, but the McMurdo is much more difficult to find. It has hand painted Perspex panels. 3" tall. $2,000-2,500.

MW & COMPANY, ca. 1932. An English-made sterling silver lighter. Possibly Mappin & Webb. 2" tall. $200-250.

MCMURDO, ca. 1950. A very rare Perspex aquarium table lighter from the British manufacturer McMurdo. Similar to the Dunhill version, but with the patented McMurdo action. $2,000-2,500.

MOSDA, ca. 1933. A made in England, leather wrapped lighter that uses a Colibri-like kick-start mechanism. 2" x 1.3". $200-250.

NETOR, ca. 1934. A most unusual silver and chromium plated brass semi-automatic lighter. Push the button on the right and both the cap and lift arm spring open, igniting the flame. Inside the lighter is marked "THE NETOR BRITISH MADE" and "PRO[VISIONAL] PAT 17391." The personalization is quite interesting. By adding a few spaces you get "Diddle Us from Diddles Us and Diddles Us." "Did" is a word with a variety of definitions, including "cheating" and "sexual intercourse." $600-700.

ORA, ca. 1934. A sterling silver lighter made in Birmingham England in 1934. It has a manual operation. $150-200.

ORLIK, ca. 1926. England. Three Orlik pocket lighters with windguards (note that each windguard is different). The model on the left is covered in Shagreen and is marked as follows: "Pat 124760 Orlik 1926 Reg. 718331." The center model is engine turned metal, and is marked "Orlik Matchless London Made Reg. 718331." The model on the right is covered in snakeskin and bears the mark "Orlik 144 London Made Reg. 718331." *Left and right:* $100-150 each; *center:* $75-100.

ORLIK, ca. 1926. A pair of English made Shagreen and ostrich covered lift arm lighters. $60-90 each.

ORLIK, ca. 1930. England. A very unusual manual lighter with a semi-automatic mechanism. Nickel-plated brass with leather wrap. 2" tall. $500-800.

ORLIK, ca. 1926. England. A sterling silver watch lighter with a windguard. Although not marked "Orlik," it gives every indication of being made by Orlik. 2.2" x 1.7". $2,500-3,000.

ORLIK, ca. 1926 England. Pocket lighter with windguard. Very rare sterling silver lighter with concealed mirror on the side opposite the one shown. Hallmarked Birmingham 1926. The elaborate engine turning surrounds the entire letter. On the other side there is a well engineered, concealed flap which lifts to reveal the mirror. Marked on bottom: "[illegible], ORLIK [model] 1926, Rd. No. 718331." $1,500-1,750.

ORLIK, ca. 1939. A salesman's case of MIB Orlik Storm proof lighters – One table lighter and ten pocket versions in a variety of finishes. The case is covered in Morocco leather. The logo of the old judge smoking a pipe is proprietary to Orlik who also made quality pipes. This case set dates from before WW II. $1,250-1,500.

PARKER, ca. 1925. An English made Parker "Beacon" lift arm lighter with an engine turned pattern. Silver plated brass finish. 2" tall. $125-175.

PARKER, ca. 1933. England. Parker "Beacon" model lift arm sterling silver and cobalt blue enamel. Parker Pipe Ltd. was founded by Alfred Dunhill in 1923. 2" x 1.4". $500-600.

PARKER, 1928. A Swiss-made Parker "Beacon" watch lighter. Made of solid sterling silver with engine turning. An extremely rare lighter with high-quality manufacturing. This model can also be found in silver-plated brass and with a round watch. 2" tall. $1,600-2,000.

PARKER, c.1935. Parker "Beacon" small size lift-arm lighter with windscreen.. This particular model was manufactured in Switzerland, but was sold in France as it bears a French lighter tax stamp. Chrome-plated brass with engine turning typical of the Art Deco period. 1.75" tall. $175-250.

PARKER, ca. 1929. A "Beveled Oblique" (diagonal cut sides) lift arm 9kt gold sleeve over brass. 1.8" x 1.3". $400-500.

PARKER, 1937. A British-made "Parker Beacon" lift-arm lighter in solid sterling silver, with a rarely seen basse taille ornamentation on the front and a Royal Air Force winged insignia. 2" tall. $1,000-1,200.

PARKER, ca. 1929. A small size lift-arm in sterling silver with a gold wash and another in sterling silver with a pale lavender guilloche sleeve. 1.3" x 1.4". $300-500.

PHILIPS, ca. 1950. England. Unusual light bulb shaped lighter promoting the Philips Company of Holland. $50-75.

PLANET, ca. 1940. English made, brass trench type lift arm lighter with an unusual top fluid fill. 2" x 1.5". $125-175.

POLO, ca. 1939. England. Chrome plated lift arm with a double-wheel mechanism. Pretty Art Deco engine-turned pattern. $30-50.

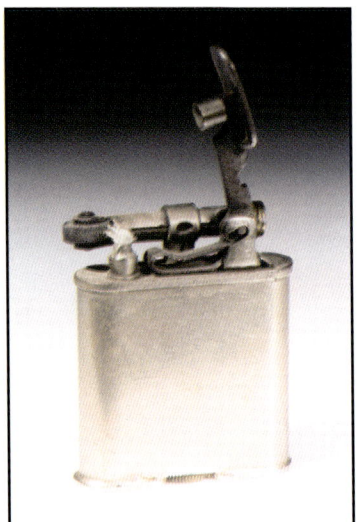

POLO, ca. 1934. England. Chrome single wheel lift arm with unusual snuffer cam. $30-50.

POLO, ca. 1945. England. Chrome column shaped table lighter on black bakelite base. 6" h. $50-75.

PREMIER, ca. 1927. A chrome plated table lighter, made in England. It has a well recognized semi-automatic, lift arm design seen more commonly in the pocket version. 4" x 3.2". $350-400.

UNITY, ca. 1935. Glasgow, Scotland. Sterling silver with a pink guilloche finish. 2" x 1.1". $400-500.

ROSE, ca. 1940. A sterling silver roller bar lighter with gold accents. Made in England. $400-500.

UNITY, ca. 1935. Glasgow, Scotland. Sterling silver semi-automatic lighter with a black lacquer and blue guilloche finish. This lighter has no visible screws and is re-flinted and fueled by pulling the snuffer cap and assembly from the body. 2.4" x 1.3". $400-500.

ROBERT LEWIS, ca. 1949. The lighter on the left is silver plated and bears no markings of any sort, but its mechanism is identical to the one on the right. It is also the same size. The model on the right is done in sterling. On the bottom it is marked "Robert Lewis" and "Made in England." The lighter, screw, and windguard are all marked for London. The date hallmark is 1949. This is a very well crafted lighter. As you can see, one can raise the windguard after igniting the wick. Most likely, that function was used without lighting the wick, but as a step in replacing the flint. There is no flint screw. There are two mall screws on the front top side of the lighter just beyond the flint tensioning wheel. When the windguard is in the lifted position, those screws are removed, the tensioning wheel can be unscrewed, and the flint replaced. A complicated bit of machinery. $750-1,000 each.

UNITY, ca. 1936. Glasgow, Scotland. Rare table lighter in sterling silver. 3.2" x 1.3". $400-500.

Unknown, ca. 1916. Fully hallmarked, English sterling silver pocket striker lighter. $150-200.

Unknown, ca. 1922. An English-made fire extinguisher shaped table lighter with lighter inside. $100-200.

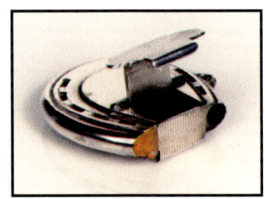

Unknown, ca. 1918. An English made, sterling silver match vesta with a tinder wick. 2.25" tall. $250-350.

Unknown, ca. 1927. Sterling silver octagonal lift cap manual lighter with windguard. Very fine craftsmanship. Windguard is hinged at the rear. Nicely engine turned and engraved. Hallmarked as an import to Britain in 1927. Also bears a maker's mark "C&C." $250-350.

Unknown, ca. 1922. English made, sterling silver manual lighter that is hinged on the wide side with an unusual top fluid fill. 2" x 1.5". $250-300.

Unknown, ca. 1929. England (surmised). Catalin-red colored semi-automatic lighter. Base screw is marked "Foreign Made." 2.4" x 1.5". $175-275.

Unknown, ca. 1929. England. Wonderful graphics on this early lithographed metal sign advertising "Will's Gold Flake" cigarettes. Notice the Colibri "Original" lighter.

Unknown, ca. 1932. An English made 9kt gold single wheel manual lighter with a pivoting cap. The jeweler's mark says "AWR." This has the look of an Everest lighter. 1.5" x 1.2". $750-1,000.

Unknown, ca. 1929. Early tobacco related postcard.

Unknown, ca. 1935. An English striker lighter made of sterling silver. It has an interesting shape. $125-175.

Unknown, ca. 1929. Postcard with Valentine's Day match box. The black horizontal band is an actual striking surface on this card.

Unknown, ca. 1936. England. Art Deco chrome table lighter. Flick the tail of the airplane and it lights. No markings. $350-450.

Unknown, ca. 1930. England. Unusual side arm, lift cap manual lighter. Marked on the bottom with three small hearts with the letters "P & G" respectively. Also marked "sterling silver case" and U.S. Pat. 1022140." This patent dates from 1912 and is for a very different looking design, but with a flint arm mechanism. $300-400.

Unknown, ca. 1938. Sterling silver and glass enamel lighter and case with a second similar lighter. Hallmarked as being made in Birmingham, England in 1938. Lighter, $400-600; Set, $800-1,000.

Unknown, ca. 1939. An English-made chrome striker lighter on a black base. It stands 3.5" high. $150-200.

Unknown, ca. 1939. England. Simply lift the cap (hinged on the left side) and rotate the flint wheel to ignite the wick. Advertiser for "Ascot Gas Water Heaters Ltd." Marked "British Made" and "British Reg. Design No. 823691." $100-150.

Unknown, ca. 1940. England. Interesting plastic, book shaped lighter. $100-150.

Unknown, ca. 1950. Probably an AD Bach, English made lighter. Hallmarked sterling silver, with a side thumb roller and a bottom slide for refueling. $250-350.

Unknown, ca. 1940. A pair of very unusual looking plastic lighters with hinged lids. The red one is about 4" tall and the white and black one is about 3" tall. Both were probably made in England. $100-150.

VICEROY, ca. 1928. A group of English made 9kt gold and sterling silver single wheel manual lighters. Model on the right is gold while other two are silver. The smallest is 1.3" x 1.1". *Left:* $300-400; *center:* $300-400; *right:* $600-800.

W & G, ca. 1912. A brass lighter with horse motif. $50-75.

Germany

BRUMA, c.1925. A very rare German lighter in solid silver. Engine turning and "935" silver hallmark. 2.2" tall. $800-1,000.

AUGUSTA-ZUNDER, ca. 1942. German made chrome automatic lighter engraved "Kiel." $40-60.

BRUMA, ca. 1935. Selecta SB Germany. A rare and highly collectible lighter with an interesting plunger mechanism. $350-450.

AUTI, ca. 1939. A nice, German made, black enamel table lighter with exposed alpacca stripe. Shown with a pocket version. Size of table lighter is 5" x 2.8". Left: $1,750-2,000; right: $250-350.

BAIER, ca. 1946. A very high quality, small rolling table or cart lighter, ashtray, and cigarette case. All metal (aluminum and brass) except for the tires. Made in Germany. $200-300.

BRUMA, ca. 1930. All of these lighters use the same well designed, semi-automatic mechanism. The flint wheel has a section removed (about 1/3). The push button is designed to catch the notch in the wheel and manually rotate the wheel while simultaneously lifting the cap. All markings are on the fluid screw unless otherwise indicated. All are marked with the same star seen on the windguard of the one in the center. Now for the differences: Only the two versions on the right have identical windguards. The plunger of the first model shown on the left is unique in its hexagonal shape and stylized snuffer. It is marked on the bottom "Bruma," "Selectra," and "SB" (probably S. Bruckman). The model second to left has a lovely windguard in the shape of a lizard. It is also marked "Bruma" and "SP," but also "D.R.G.M." It also has a small rectangular stamp of "1935" on the guard. Next is the center model, which is silver plated, and marked "Press Easy," "Germany," "Pat. Appl. For," and "Platinin." The next model is leather covered and marked simply "PATENT APPLIED FOR." Finally, on the right we have the small leather covered model marked "Made in Germany," "U.S. Pat. Pend.," and on the top of the push button "Press It Lighter." Bruma is a contraction of the name "Bruckman," a well regarded German manufacturer of the time. $400-500 each.

CARLTON, ca. 1931. A German automatic, chrome plated lighter with an attached French metal tax stamp on the right side. 2" x 1.6". $150-200.

DUNHILL (FAKE?), ca. 1929. On the left we see a fake Dunhill lighter with a semi-automatic lift arm. There is no record of such a lighter being made, even in prototype. Moreover, this lighter bears all of the markings on the bottom that would have prepared the model to be sold at retail, which further suggests this lighter is not a prototype. The mechanism is activated by pushing a small button on the other side of the notched wheel. The spring (wrapped around the "axle" between the notched wheel and the opposite screw) causes the arm to lift. The wheel itself is still manual. Note the similar mechanism on the small lighter to the right. It bears German markings.

DIKSI, ca. 1928. Germany. A 14kt gold straight lift arm, double wheel lighter with detailed vertical engine turning. (It is missing the thumb wheel that would be located below the spark wheel.) 1.8" x 1.5". $500-750.

DURKOPP, ca. 1918. A German striker lighter in sterling silver with an unusual triangular shape. $150-200.

DOBEREINER, ca. 1830s. A heavy piece. Probably the top to an old Dobereiner lamp, about 5" in diameter. The Dobereiner lamp was invented in the early 1820s by Johann Wolfgang Dobereiner. It was a catalytic lighter where sulfuric acid was contained in a glass jar and a zinc rod was pushed into the acid. This created hydrogen gas which, when exposed to a platinum wire, created a flame. Also produced in Austria. $1,500-2,000, if complete.

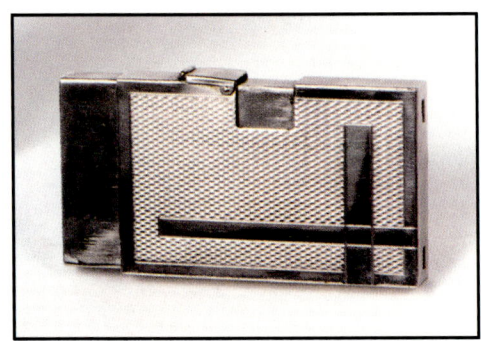

DURUS, ca. 1939. A German, chrome plated center opening lighter that is squeezed to activate. 2.5" x 1.4". $300-400.

ECLYDO, ca. 1950. This German-made watch lighter had a novel winding mechanism. Each time you depressed the lighter mechanism it wound the watch. Similar systems are sometimes found on musical lighters where each lighting movement activates the music box. 2.5" tall. $150-200.

E.D. (EMIL DEUTSCH), ca. 1911. Well made and engineered semi-automatic lighter in the form of a pocket watch. The mechanism is a gear and spring action. It's a variation on the RK mechanism (which has two gears at the flint mechanism, while this one has only one, the second one being replaced by a watch type spring). Push the "watch stem" to activate. Note the small thumb catch protruding under the stem. This opens the other side of the case to reveal the mechanism. Each piece is precise. The lighter bears a number markings. On the front of the internal mechanism we see "D.R.G.M. GES. GESCH." indicating German origin. Then "PATENT ANG." indicating British patent protection. Then what appears to be a maker's mark "ED." On the reverse side of the flint retainer is a clearer version of the ED mark which looks like this within a box: E*/D, actually the asterisk has a stem between the two bottom points. Also on the mechanism we see "REGD No. 573341" and "DEPOSE" indicating France. Finally there is a very clear "H" on another part. $400-600.

HAHWAY, ca. 1916. A great sterling silver semi-automatic action lighter. 2.5" tall. $250-300.

HENCKLES, ca. 1933. German. Elaborate Smoker's companion with fitted ashtrays, match stick holder and striker, cigar cutter (the propeller) and nesting ashtrays. The body serves to store cigars, and the removable wings as cigarette cases. When the cigar cutter is used, the droppings fall through a chute into the piece to which the wheels attach. It is removable so it can be emptied. This is of the same manufacture as the well known drinker's companion/martini shaker of the same period. $2,500-3,000.

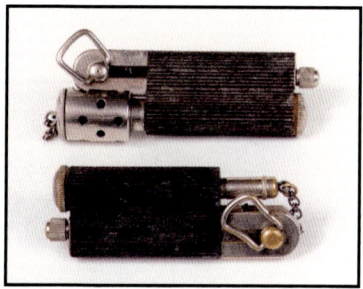

KASCHIE, ca. 1922. Model "8" key turn lighters in chrome plated pot metal. Two models shown both with and without a windguard. The windguard version is rare. Made in Germany. $250-350 each.

KASCHIE, ca. 1942. Chromium plated, lever action automatic lighter. Bottom has an inset which slides out to allow for refueling. $75-100.

KASCHIE, ca. 1945. A pair of Model 38's. A very Art Deco looking, German made, automatic lighter with elaborate enameling with gold and chrome plating over brass. Both have the customary cross mark cut out on the front. The red and gold model has the visual fluid lever check slot. The black and gold one does not. 1.9" x 1.4". $150-200 each.

KW, ca. 1933. Two German "KW" lighters with sharkskin (shagreen) coverings. $100-125 each.

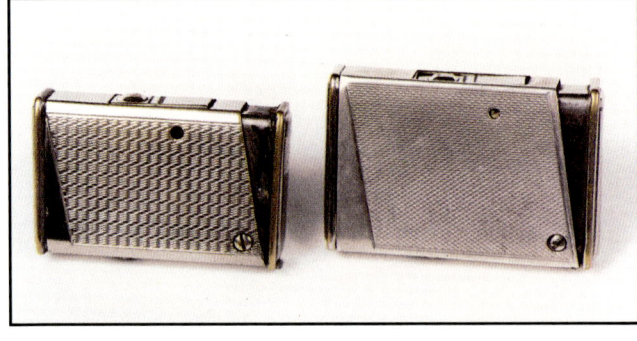

KW, ca. 1942. On the right is a 700 silver center opening lighter. The lighter is activated by pushing in the right side. The lighter on the left is just marked "Germany." *Right:* 2.1" x 1.5"; *left:* 2" x 1.3". $400-500.

KW, ca. 1939. Two lighters with identical mechanisms, semi-automatic. Both are encased in sterling sleeves. Both marked with the German crown and moon hallmark and .800 silver. The table lighter bears the maker's mark "SS." $100-200 each.

NEGBAUR, ca. 1941. The "Dome" lighter with its box. Very similar to the American made "Speed" lighter. 2.25" tall. $100-125.

KW, ca. 1940. German made lighter, mint in the box. Chrome plated with a blue celluloid wrap. $80-120.

PIONEER, ca. 1932. A German made lift arm model with a center wheel and a blue glass enamel sleeve. 1.9" x 1.6". $250-350.

ROYAL, ca. 1950. German made "Gussing" automatic lighter. $30-50.

SARASTRO, ca. 1928. The rear lighter is a large sport model made of alpacca (German word for non-precious metal). It is nicely engine-turned. It sports the signature Sarastro windguard. Lift the arm and find a well-engineered thumb wheel, machined into the body of the lighter just below the windscreen (not visible from this side). The only markings appear on the fuel screw: "SARASTRO, ALPACCA, DIKSI-SPORT." The 14k gold version in the front is beautifully engine turned and bears only one mark. On the bottom is ".585" indicating the gold content. It certainly appears to be a Sarastro, given the signature windscreen. However, it is a single wheel version, meaning you must thumb the actual flint wheel, rather than the more tasteful mechanism in the alpacca version. Perhaps this is just an indication that the gold version is from an earlier date. *Rear (alpaca)*: $100-200; *front (gold)*: $500-750.

SARASTRO, ca. 1928. Germany. Marked on the fuel screw "Diksi Sport" model with windguard. Fixed windguard. It has the signature Sarastro windguard and flint wheel mechanism. $200-250.

TACO-LITE, ca. 1924. A German-made leather lighter that is similar to a French Lancel lighter. 2" tall. $125-150.

Unknown, ca. 1912. A very unusual combination cigarette lighter/shoe horn. This was made in Germany and most likely combined after production. $100-125.

SARASTRO, ca. 1928. Germany. The model is "Diable" marked on the fuel screw, which also bears: "SARASTRO," "ALPACCA." This version is unusual in several respects. First, the flint tube, flint wheel, and windguard are formed into a single unit. Second, the flint wheel mechanism differs radically from the customary Sarastro model. Third, although this lighter is made of Alpacca, it is marked on top with a makers mark "B L" in a rectilinear octagonal box; perhaps this was a prototype of which the maker was proud. About the only thing reminiscent of the standard Sarastro sports model is the cut out design of the windguard itself. $250-500.

Unknown, ca. 1914. Germany. Two identical Catalin table lighters, each with a different coloration. Both are marked "D.G.M." on the thumb piece. The piece which juts from the top is the flint tension mechanism. Before lighting, you must first manually lift the wick cap. The entire bottom screws out to allow for refueling. $150-250 each.

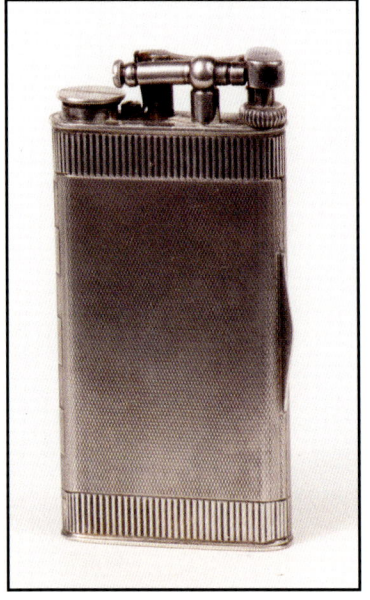

Unknown, ca. 1924. Sterling silver combination cigarette case and curved, lift arm lighter. Appears to have been made by the same craftsman in Germany who also did these types of Dunhill lighters under contract.

Unknown, ca. 1916. Probably German or Austrian. Small brass figural pocket striker in the form of a violin. The finial at the top comes out and is attached to the wick housing. The striking surface is on the lower part of the face of the violin. $200-250.

Unknown, ca. 1926. A German, sterling silver and blue enamel lighter with a watch. The front opens down from the top to access the watch. $3,000-4,000.

Unknown, ca. 1922. A German made striker lighter in the form of a motorcycle with two riders It is marked "DRGM" indicating that it was made in Germany. $200-300.

Unknown, ca. 1923. A German made sterling silver and enamel lift arm lighter with a beautiful pastoral scene. The bottom is marked "Sterling Germany." 1.8" x 1.5". $1,000-1,500.

Unknown, ca. 1926. A German lift arm watch lighter in a hard glass enamel. It has a round watch and beautiful yellow with red numerals 13 through 24 around the face of the watch. 2.1" x 1.5". $5,000-6,000.

Unknown, ca. 1928. A German lift arm watch lighter in a green hard glass enamel and chased sterling silver. It has a round Swiss watch. 2.1" x 1.5". $3,000-4,000.

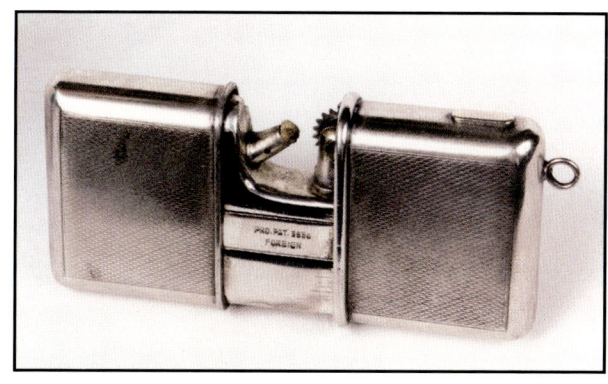

Unknown, ca. 1930. A silver, center opening, automatic slide lighter. Believed to be made in Germany. 2.2" x 1.5". $300-400.

Unknown, c.1928. A small lift-arm lighter probably made in Germany or Austria. Gold plated brass, hand-engraved sterling silver wrap with hard glass enamel and a miniature hand-painted scene. 1.3" tall. $350-450.

Unknown, ca. 1940. A German made, brass and bakelite lighter that was made to look like a notebook. 2.25" tall. $100-150.

Unknown, ca. 1928. A German lift arm watch lighter in a blue hard glass enamel and chased sterling silver. It has a rectangular Swiss watch. 2.1" x 1.5". $3,000-4,000.

Unknown, ca. 1944. This is not a lighter. It is a pencil sharpener. The bottom pulls out to empty the trimmings. The lift arm and flint screw are fixed. Deco enamel. Marked on the bottom "Germany". $25-50.

Unknown, c.1928. A lift-arm lighter probably made in Germany or Austria. Solid sterling silver with a miniature hand-painted scene on the front and basse taille on the sides and back. 1.75" tall. $500-700.

VICTORIA, ca. 1892. A cap lighter that uses a roll of caps, but the material which holds the caps is a waxed paper, not the typical paper. The lighter is missing the lever for triggering the cap action. Chrome plated brass, engine turned and marked "Patented Made in Germany." Shown with a roll of caps. 2.2" x 1.2". $300-400.

Austria

FIREMASTER, ca. 1938. Austria. A rare and unusual pipe lighter. Chromium plated brass. The fuel tank and wick are in the center section. The lift cap and windguard, on the left, is hinged at the bottom. This feature is used to re-flint the lighter as well as to put it into pipe mode. The flint wheel mechanism, on the right, must be slightly turned to release the cap (which must be manually lifted to expose the wick). This mechanism also folds back once the lighter is lit. By folding both sides back you get a pipe lighter. Fold back neither and you have a windproof cigarette lighter. Marked on one side "Firemaster: (in script) and "MADE IN AUSTRIA." On the bottom it is marked "PATENTS." $1,250-1,500.

A.D., ca. 1930. Austria. A 9kt gold sleeve semi-automatic flip top lighter with a short snuffer cap. Beautiful engine turning and a wonderful snap back lighting mechanism 2.2" x 1.8". $250-300.

IMCO, ca. 1933. A trench art looking simple mechanism housed in a figural brass barrel. $100-150.

CONDOR, ca. 1932. An Austrian made pull-apart, flint wheel mechanism lighter. To activate, you pulled it out of the sleeve. 2.3" x 1.3". $550-750.

JUWEL, ca. 1911. Probably Austrian or German. An unusual semi-automatic rasp mechanism lighter. Slide the button on the side down to release the spring-loaded top and the spring loaded rasp jumps up to spark the wick. Simply close the top to reload the mechanism. Iron with brass screws. Side mounted fluid screw. 2.1" x 1.7". $250-350.

FACKEL, ca. 1922. Austria. This lighter has an unusual pull out wand mechanism. A screw like wand is pulled out of its slot, drawing across flint material and a gear wheel. The top of the lighter bears a six pointed star-shaped hole which cuts through the lighter, whose purpose is to allow air to reach the mechanism and assist in ignition. A circular version was also produced. 2.3" x 1.8". $200-250.

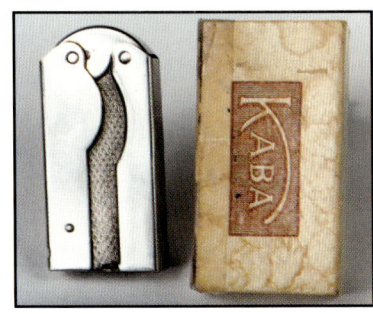

KABA, ca. 1941. An Austrian made, chrome plated lighter with a squeeze action and an unusual design. 2.6" x 1.2". $350-500.

MEB, ca. 1930. Austria. The dog's head lifts up to access the lighter. Filled through bottom opening. Not an insert. Also made as a wolf and a dachshund lighter. Importers ME Bernhardt NYC. 2" tall. $100-150.

MIROLITE, ca. 1940. Austria. A chrome plated, semi-automatic lighter. It has an oval mirror on one side and a stand on the other. It is marked Austria on the stand and "F.C.L." after the name Mirolite. A similar version exists known as the "Admira" but with a shorter stand. 2.4" x 1.3". $750-1,000.

ORLIK, ca. 1933. Austria. A semi-automatic Orlik with a lever mechanism (lever is on reverse side shown). The three holes are not for wind protection. On the contrary they are intended to increase the air flow from under the wick to create a more robust flame. Both sides are identical, absent the lever on one side only, bearing the same engine turned designed and Art Deco relief and initial plate. This lighter refuels by sliding the bottom out of the lighter which reveals the cotton wadding and the flinting mechanism. Marked "PAT. PDG, ORLIK MAYFAIR, AUSTRIA." $500-750.

NOBLESSE, ca. 1953. A pair of Austrian enameled semi-automatic wind proof lighters. Nickel plated brass and black enamel, each with a different view of a terrier with a colorful collar and leash. 2.1" x 1.5". $100-150 each.

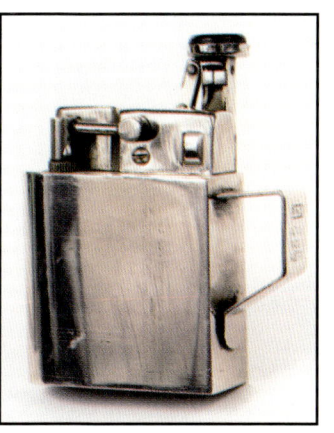

PINO, ca. 1935. Austrian table lighter. Chrome plated brass with black plastic push button. Marked Pino and Made in Austria. A manual version was also produced Also has a crown engraved with the letters k i n g interspersed between the prongs. 3.5" x 3". $200-250.

ORLIK, ca. 1925. Austria. Very rare pocket light with side to side juxtaposition of the flint wheel and the wick. Nickel plated brass with no markings on the base. The side not shown bears very clear markings: "ORLIK 1925, MADE IN AUSTRIA." $150-250.

R. K., ca. 1910. (R. K. stands for Richard Kohn.) A sterling silver semi-automatic lighter with elaborate engraving. Hallmarked for Birmingham, England. 2.5" x 1.4". $300-400.

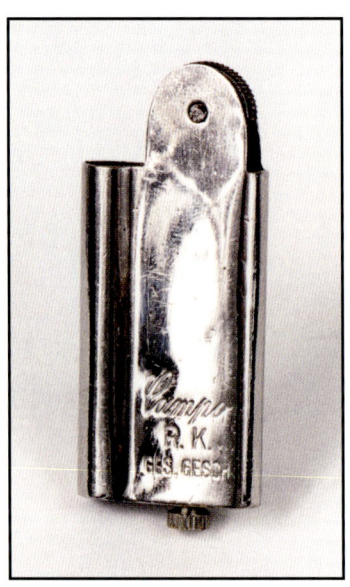

R. K., ca. 1910. Austria. (R. K. stands for Richard Kohn.) A primitive and simple tinder lighter with exposed large flint wheel and flinting screw. Marked R. K. GES. GESCH. Also marked LAMPE which means "lighted lamp" which was an early term for lighter. 2.2" x .9". $250-300.

TCW, ca. 1920. Treibacher Chemische Werke. An Austrian Taifun Early Zippo-type windguard model many years before Zippo ever existed. This is a chrome plated lighter with a brass insert and electrical advertisement. $60-90.

RK, ca. 1918. Austria. A 14kt gold woven tinder wheel lighter. The button to release the lid is set with a cabochon ruby. The piece is hallmarked for Louis Kuppenheim, a very fine German silver and goldsmith. This was one of his earliest pieces. This lighter did not use fluid; the sparks caused the tinder cord to smolder without a flame. The exposed wheel near the base was used to roll the cord up or down. 2" x 1.3". $5,000-6,000.

Unknown, ca. 1908. Two pocket strikers The one on the left is sterling with a hard glass enamel light blue over an engine turned sunburst. Marked .925 (sterling silver) and hallmarked as being imported into Great Britain. The one on the right is sterling with dark blue enamel decorative vertical design and still has its original flint material. There is a hallmark but it is not legible. The lighter appears to be Austrian. *Left:* 1.9" x 1.6"; *right:* 1.8" x 1.1". $250-350 each.

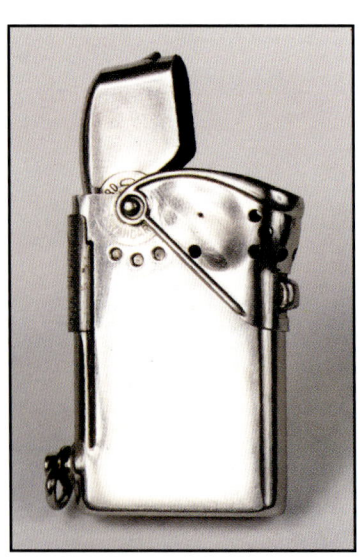

STANDARD, ca. 1922. An Austrian made, semi-automatic chrome plated brass lighter with a windguard (which can be raised out of the way). It has a side fluid screw with a ring for a chain and slide access to the mechanism for re-flinting. Re-flinting slide marked "TO EXCHANGE STONE PUSH SLIDE DOWN." An elaborate arrow shows the direction in which the slide moves. There is a diamond shaped box with MEB (the U. S. importer) inscribed inside. Very rare version. 2.3" x 1.5". $150-250.

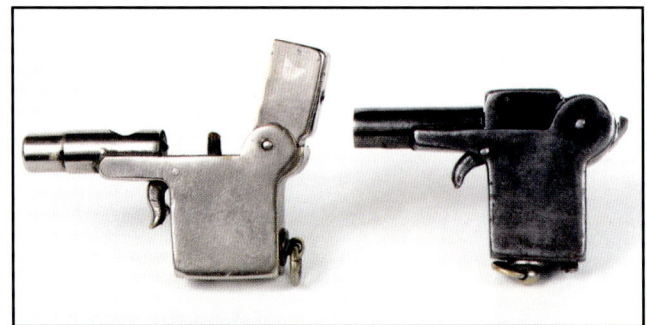

Unknown, ca. 1912. Austria. A pair of figural gun, semi-automatic, flint and petrol lighters. The one on the left has a built in hidden cigar cutter (the extended barrel is actually part of a cutter device which is more visible when viewed from the top). Both the iron version (on the right) and the plated version (on the left) have brass rings for attaching to a fob or watch chain. In both models the flint screws are capable of being screwed in flush with the bottom of the handle. Possibly Austrian "Rival" lighters. 1.5" x 2.4". *Left (with cutter):* $400-500; *right (without cutter):* $300-400.

Unknown, ca. 1918. An Austrian made, RK type lighter in .800 silver with a great embossed rooster design. 2.3" x 1.4". $500-600.

Unknown, ca. 1919. A three piece Austrian set lighter, cigarette case, and ashtray in a gold plate with orange hard glass enamel. $350-450.

Unknown, ca. 1922. Austrian striker lighter in the shape of a dog with pipe rest. Hinged back opens to reveal a cigarette rest and ash receptacle. Also seen as a Scottie dog. $200-300.

Unknown, ca. 1920. A European striker lighter, believed to be from Austria, in the form of a heavily petticoated woman. The striker rod is located on her upper left side and the striker strip is just below it. Seen in many color combinations and with the rod placed in other locations. $200-250.

Unknown, ca. 1922. Austrian striker lighter in form of camel and magic carpet. Referred to as "The Carpet Seller." 8" h. x 10" l. $250-300.

Unknown, ca. 1922. A beautiful enameled scene on both sides of this European lift arm lighter make is desirable. The lower body is made of silver. Made in Austria or Germany. $700-900.

Unknown, ca. 1923. Austrian striker in the form of a bird on a branch with the striker in the tree trunk. The use of the tree trunk as a wand receptacle was a common technique with figural strikers. 7" l. $175-225.

Unknown, ca. 1924. An Austrian, lift arm sterling silver model with a translucent green enamel. $800-1,200.

Unknown, ca. 1928. An Austrian lift arm lighter with a watch, with blue, hard glass enameling. 1.8" x 1.6". $3,500-4,000.

Unknown, ca. 1924. An Austrian made, striker lighter with an Indian. 8" h. $350-550.

Unknown, ca. 1928. An pair of similar enameled lighters in sterling silver. On the left, Germany, red opaque enamel. $500-$700. On the right, Austria, translucent green enamel. $800-$1000.

Unknown, c.1926. Austria (surmised). A very rare watch lift-arm lighter in solid sterling silver, with hand-engraving and hard glass enamel. 1.8" tall. $1,800-2,100.

Unknown, ca. 1928. Another pair of similar enameled lighters in sterling silver. On the left, Austria, multi-colored translucent enamel. $800-$1,000. On the right, Germany, very similar to a "Kulturstaat" with its button on top. $800-$1000.

Unknown, ca. 1926. Sterling silver and green, Austrian made, enamel lighter with a watch. The front hinges open for access to the watch. $4,000-5,000.

Unknown, ca. 1929. Beautifully engraved 14k (marked on top) lift arm lighter. Not only is the body engraved, but also the top and sides of the lift arm cap. Both the bottom and the fuel screw are also engraved. The bottom also bears a stamp ".585," being the European mark for 14kt gold. From the quality and artisanship, probably Austrian or German. $1,000-1500.

Unknown, ca. 1933. An Austrian made hunting dog striker lighter. The wand and flint are located on the tree stump. The technique of using the stump as a receptacle for the wand is seen on many of the Austrian striker lighters. 10" l. $300-400.

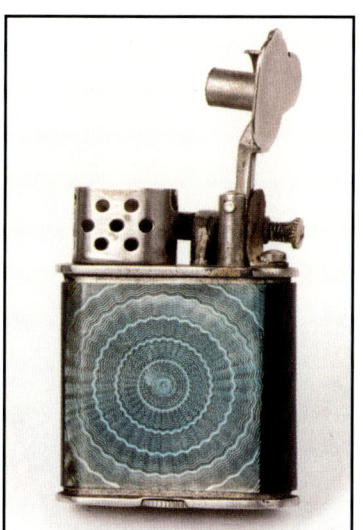
Unknown, ca. 1929. Lift arm lighter with a fixed windguard. Some sort of plastic-like wrap imitating hard glass enamel. Judging from the chips on the side, this lighter was first sleeved with the green product and then the sides were simply painted black over the green. The screw is marked Austria and bears an octagonal mark which is unknown to us. The lighter itself is of decent quality nickel plated brass. The wrap was probably done later. $125-150.

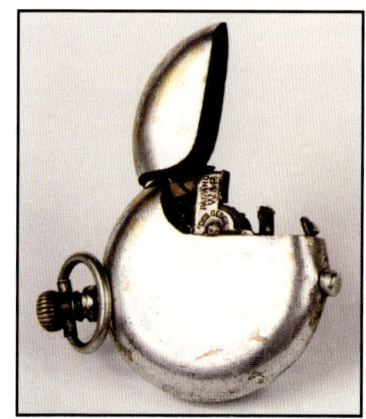
W & P, ca. 1918. Austria. A pocket watch form, semi-automatic, push button, gear driven, spring loaded mechanism. The "watch stem" serves as the fluid screw. A separate button on the bottom releases the mechanism. The mechanism is marked "PAT. ANG. W.& P. GES GESCH." 1.8" x 1.8". $300-400.

Unknown, ca. 1929. A pair of Austrian lift arm lighters, done in green enamel. The one on the left has a lady's compact while the one on the right has a watch. 1.8" x 1.6". *Left:* $4,000-5,000; *right:* $4,000-5,000.

Italy

Unknown, ca. 1953. Italy. Pretty hard glass enameled case with Zippo insert. We believe these Italian made cases were a joint lighter venture between some of the Italian silversmiths and Zippo Mfg. Co. some fifty years ago. $150-200.

Unknown, Italy (surmised). 17th century split barrel tinder pistol. A well-made fire-making device. Upon impact of steel upon flint, a pinch of gunpowder is ignited which lights a candle concealed within the barrel. The spring-loaded candle rises to illuminate the darkness. 10" x 4.8". $5,000-7,500.

Unknown, ca.1954. Italy. Beautifully enameled scene with man and woman. The insert is marked Zippo. 150-200.

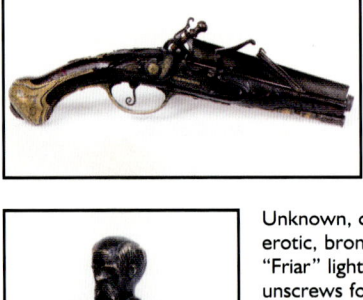

Unknown, ca. 1922. An erotic, bronze, figural "Friar" lighter. The bottom unscrews for refueling. Lift his arm to reveal wick. Made in Italy. $400-500.

Unknown, ca. 1954. Italy. Beautifully enameled courtesan scene. The insert is marked Zippo. $150-200.

Unknown, ca. 1953. Italy. Beautifully enameled courtesan scene. Insert marked Zippo. $150-200.

Unknown ca. 1954 Italy. Beautifully enameled with woman and goat. The insert is marked Zippo. $150-200.

Switzerland

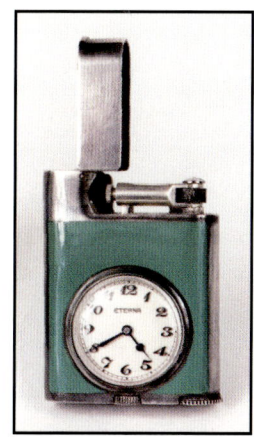

ETERNA, ca. 1929. Switzerland. A watch lighter in sterling silver with a green enamel finish. It has a round watch face. 2.3" x 1.5". $4,000-5,000.

COMBO, ca. 1939. Three lighters – the ones on the left and right are sterling silver windguard models. The center lighter, with black lacquer finish is marked Dunhill. $500-700; Dunhill model $1,500-2,000.

ETERNA, ca. 1929. Switzerland. Two wonderful watch lighters. On the left is a sterling silver model with floral enamel. On the right is an engine-turned sterling silver in ladies size. 1.875" h. Left: $2,500-3,000; right: $1,500-2,000.

ETERNA, ca. 1929. Switzerland. This lighter is in sterling silver and has a rectangular watch. 1.9" x 1.4". $2,500-3,000.

ETERNA, ca. 1929. Switzerland. A watch lighter in sterling silver with a blue hard glass enamel incorporating a flower design front and back. It has a round watch face. 1.8" x 1.4". $5,000-6,000.

ETERNA, ca. 1930. A pair of superb watch lighters in sterling silver. Both have outstanding Art Deco enamel work, round watches, and were made in Switzerland. The front of the lighter on the left has plain, vertical yellow lines, while the back has two colorful birds. There is evidence of a gold wash. The lighter on the right has an Art Deco green and black enamel design with a string of Jaguars. There is evidence of gold wash. Note: Lighters in bottom photo have been reversed. 2.3" x 1.5". $7,500-10,000 each.

ETERNA, ca. 1932. A watch lighter in 14kt gold. It has a round watch and was made in Switzerland. 2.3" x 1.5". $8,000-10,000.

HERMANNS, ca. 1934. Early version of Hermanns lift arm watch lighter with a leather cover and concealed watch. 2" x 1.7". $1,250-1,500.

HERMANNS, ca. 1934. Later and much improved Hermanns lift arm watch lighter with an Art Deco design in sterling silver and black enamel. It also has a concealed watch. 2.1" x 1.6". $7,500-10,000.

JUVENIA, ca. 1929. A Swiss made lighter that is semi-automatic, nickel plated over brass, with an egg shell design on the side. It is very Art Deco. $1,250-1,500.

JUVENIA/PALL MALL, ca. 1930. A pair of Swiss watch lighters in sterling silver. The one on the left has a watch marked "Bucherer" and the one on the right has an outstanding blue translucent hard glass enamel over an engine turned body. Its watch is marked "Juvenia." 2.1" x 1.5". Left: $2,500-3,000; right: $10,000-12,500.

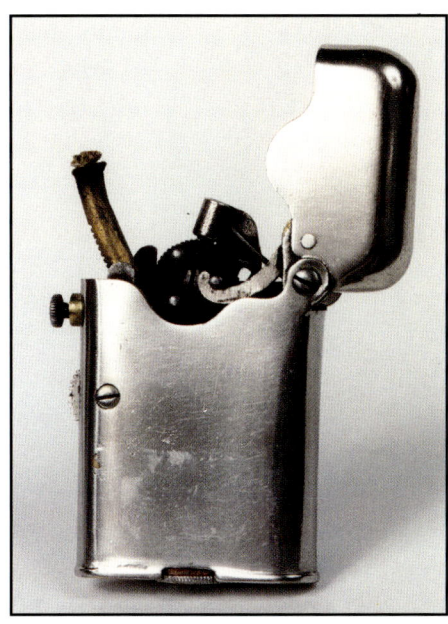

THORENS, ca. 1922. Switzerland. Automatic push button "Teleflam" pipe lighter. The wick extends by turning the wheel under the safety lock. The earliest Thorens models have a single "claw" which rotates the flint wheel. Later versions use a "double claw" mechanism which grabs the wheel from both sides. The first Thorens lighter which has a US patent date of 1914, was unique and innovative because of its flint wheel being a separately fitted piece. Chrome plated brass. $400-500.

THORENS, ca. 1932. Switzerland. A great Art Deco bronze nude holding a semi-automatic table lighter. 3.75" tall. $250-350.

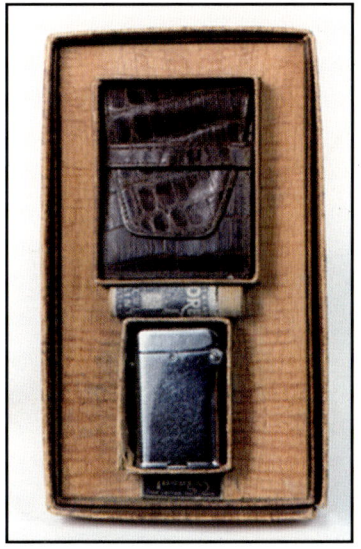

THORENS, ca. 1934. Lighter and cigarette case with spare flints in its original box. 1930s model with double-claw. $125-175.

THORENS, ca. 1924. Extremely rare if not one-of-a-kind. This Thorens has a round watch rather than the more common rectangular watch. Moreover, the watch in the more common version is mounted at the edge rather than the center of the lighter. The small indentation at the lower right of the lighter allows the bezel to pop off. By then removing the stem, the watch can be removed, unlike the screw out release button for the semi-automatic mechanism. This was a safety feature. The earliest version of this lighter had a simple button which could be easily be discharged accidentally. Marked on the bottom "THORENS," "SWISS MADE." $2,000-3,000.

THORENS, ca. 1934. Lighter fluid filler made in Switzerland. A chrome plated brass model found in shops which allowed the public to fill their lighters. Seen with both English and French wording. 5" h. $150-200.

THORENS, ca. 1936. A chrome plated, watch lighter with original box. 2.3" x 1.6". $2,000-2,500.

THORENS, ca. 1949. An extremely rare "Masterpiece" chrome plated lighter with a key wound music box. 2.5" x 1.3". $2,500-3,000.

THORENS, ca. 1939. Attractive "Oriflam" circular chrome plated model. 2" dia. $175-225.

THORENS, ca. 1949. Switzerland. A nice large (4.25" tall) table model lighter with tortoise sides and a gold plated brass finish. $90-130.

THORENS, ca. 1948. Switzerland. A gold plated "Blizzard" lighter with a windguard. It is similar to a "Masterpiece" with the exception of the windguard, and therefore much more rare. 2.3" x 1.6". $400-600.

Unknown, ca. 1911. A Swiss made, sterling silver and blue enamel pocket striker lighter with a watch. $2,000-2,500.

Japan

BROTHER-LITE, ca. 1950. Japanese flat advertising lighter promoting "Royston Tomatoes" of Bristol, Tennessee. $25-45.

BLUE BIRD, ca. 1950. A Japanese musical lighter with U. S. Marine Corp. insignia. The music box winding mechanism is on the back. $75-125.

CORNWALL, ca. 1955. Japanese flat advertising lighter for the "Skeena Hotel. $25-35.

BLUE-BIRD, ca. 1954. Japanese musical lighter commemorating the St. Louis Bi-centennial of 1954. $85-125.

CORONA, ca. 1952. Japan. A "Renown" model. A great brass lighter, automatic with a permanent windscreen. Built into the face of the lighter is a very nice slide rule in a circular form. 2.2" x 1.5". $100-150.

BRONICA, ca. 1970. The Bronica rocket ship lighter was a butane model that was made in Japan. The world below had a rotating time dial. 8" tall. $95-125.

DUNDEE, ca. 1955. Japanese flat advertising lighter for a Canadian trucking company. "Richard Transport, Inc." of Quebec and Pont Rouge. $25-35.

DUNDEE, ca. 1955 Japanese flat advertising lighter promoting the Nutty Club candy and nuts. $25-35.

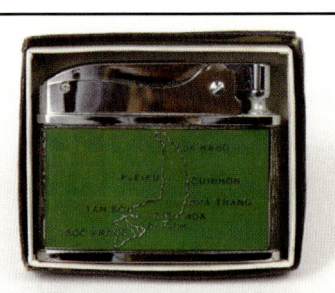

FIREFLY, ca. 1955. Japanese flat advertiser with interesting Vietnam inscription which reads Providers Club NHA Trang So Vietnam. $50-75.

DUNDEE, ca. 1955. Japanese flat advertising lighter for the Colony Radio and Television store. $25-35.

HADSON, ca. 1955. Japanese, colorful flat advertising lighter for Wondercolor Lustre-pack. $25-50.

DURALITE, ca. 1955. Interesting two-sided Japanese flat advertising lighter with the EADS Bridge, St. Louis, MO on front, and the ship S. S. Admiral on its reverse side. $30-40.

HADSON, ca. 1955. Japanese flat advertising lighter for the Carillon hotel in Miami Beach, Florida. $25-50.

IDEAL AD-LITER, ca. 1955. Japanese flat advertiser for the Mason Mint Company with a great view of the factory. $30-50.

MADISON, ca. 1955. A Japanese made flat advertiser with a World Book Encyclopedia ad. $30-50.

JAPAN, ca. 1954. Flat advertising lighter made in Japan for the American Advertising Co. of Brooklyn New York with an orange soft drink ad. $30-40.

NARUDAN, ca. 1955. Japanese flat advertising lighter promoting greyhound racing in Florida. $35-45.

NESOR, ca. 1959. A musical lighter. The real appeal of this lighter is the image of Shea Stadium on one side and the image of the New York Mets mascot on the other. $150-200.

KAY-CEE, ca. 1955. Japanese flat advertising lighter for Texaco gasoline. $40-60.

NESOR-ROSEN, ca. 1954. A Japanese made flat advertising lighter for Foremost Kosher Salami. $35-45.

NESOR-ROSEN, ca. 1955. Japanese flat advertiser for East River Savings Bank. $25-35.

NESOR-ROSEN, ca. 1955. Japanese flat advertising lighter for the Presto-lite Vibration Proof High Level Battery. $30-40.

NEW LIGHT, ca. 1950. This is an automatic pocket lighter that mint in the box, with all accessories. The model is "Zoom." It is from the 1950s and is of Japanese origin. It is very well manufactured. Press down on the upper right hand portion and the windguard withdraws (thus for decoration only) and the wick ignites. However there is a functional windguard on the reverse side, which is cleverly cut out to form the word "ZOOM." That word also appears on the top of the automatic ignition button and on the bottom of the lighter. The bottom is also marked "New Light." $750-1,000.

NEW LIGHT, ca. 1950. This is an automatic pocket lighter that mint in the box, with all accessories. This is the "Henry Automatic" model from New Light. Also from the 1950s, its design is even more complex that the Zoom. As with the zoom, there is a functional windguard (see photo) with the model name cut out. On the other side is a sliding metal door which is marked "Henry Automatic New Light." When you push up on the lever to the left, the door retracts into the housing to the left and an intense spring loaded mechanism cocks and fires to operate the flint wheel and ignite the wick. It is semi-automatic in that you must manually lower the lever to close the wick area. The other side of this lighter resembles a camera from the same era. There is a metal disk recessed into the center of the case (the covering of the case is the same as the ZOOM and closely resembles the material used to cover cameras). This disk sports a camera-like rewinding screw (pull up the flap to use). Turn it clockwise to wind the spring. Turn it counterclockwise and it comes out to allow access to the refueling area. $1,000-1,250.

PENGUIN, ca. 1964. A musical lighter advertising Gillette Razor Blades. $100-125.

PRINCE, ca. 1954. A Japanese Zippo-type lighter with attractive enamel work. 2" tall. $75-125.

ROLEX, ca. 1955. Japanese flat advertising lighter promoting Imperial gasoline. $25-35.

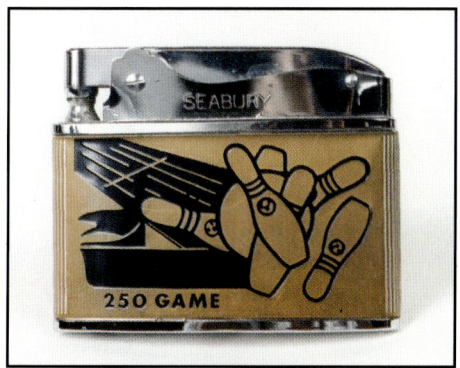
SEABURY, ca. 1955. Japanese flat advertising lighter awarded by a bowling league for a 250 game. $25-35.

SUPREME, ca. 1950. Japanese made musical lighter advertising Jell-O Dessert on the front and Reader's Digest on the reverse side. 2.5" h. $75-125.

SUPREME, ca. 1955. Japanese flat advertising lighter for Wahl Shavers. $25-50.

SW, ca. 1934. Tokyo. Japan made chrome lift arm lighter. Nicely engraved on both front and back. 2" h. $125-175.

SWANK, 1960s. Japan. Rebecca holds an outboard motor lighter which sits on a chrome and blue metal base. To operate, chrome knob on front of motor is pushed to the right. 6" h. $150-250.

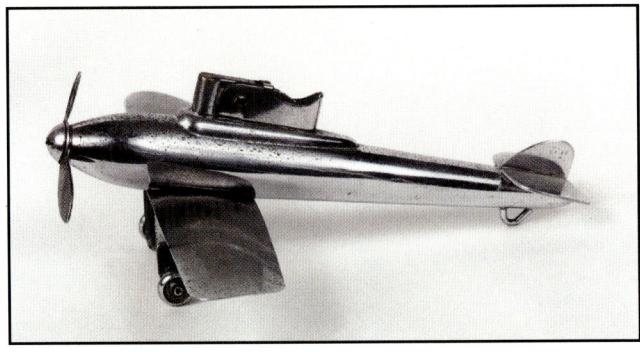

Unknown, ca. 1944. Airplane lighter that is probably made in Japan. It is chrome plated. 7" long. $200-250.

TOKYO, ca. 1944. A very unusual lift arm lighter with a spinner coin. When coin is spun the two separate faces magically appear to be kissing. $300-375.

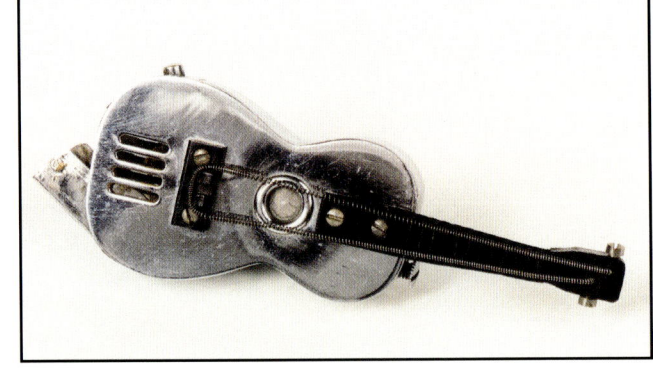

Unknown, ca. 1945. Japan. Figural chrome guitar-shaped lighter. $175-225.

Unknown, ca. 1950s. Japan made blue Sunoco gas pump. A realistic looking and beautifully enameled early gasoline pump. One piece construction and not an insert. Hinges open to reveal Zippo-type lighter inside. 4" tall. $200-250.

Unknown, ca. 1865. Two so-called walnut lighters modeled on a Japanese walnut (*Fuglans japonica*). They are called "Hiuchi's" (Tinder Box) and were hung from one's "Obi" or sash. The outer shell is made from bronze with additional use of iron, brass and/or copper. The iron ones are the earliest. They are sometimes elaborately engraved with flower appliqués, some made of silver and some of a gold/copper alloy. There are two buttons on the side and a small ring on the top (believed to be part of a system to carry the Walnut on the obi sash on one's Kimono). The larger button allows the walnut to be opened, exposing its miniature flintlock gun-like mechanism. Inside on one side there is a small bowl for tinder and a hinged striker plate, which, when struck by the flint, sparks the tinder. On the other side is the flint lock mechanism which must be cocked to activate. The smaller button on the side is then pushed, firing the device. Given the materials, this one probably dates for mid to late 19th century. *Left:* 1.4" x 1.7"; *right:* i 1.3" x 1.8". $4,500-5,500 each.

Unknown, ca. 1950. Japan. Jaguar automobile. 6" l. $75-125.

Unknown, ca. 1960. Japan. 1917 automobile (believed to be an early Mercedes Benz) with a lighter behind the driver's seat. About 7.5" l. $75-125.

VULCAN, ca. 1955. Japanese flat advertising lighter for the Stakich Furniture Company. $25-40.

WELLINGTON, ca. 1955. Japanese flat advertiser for Parliament cigarettes. $20-30.

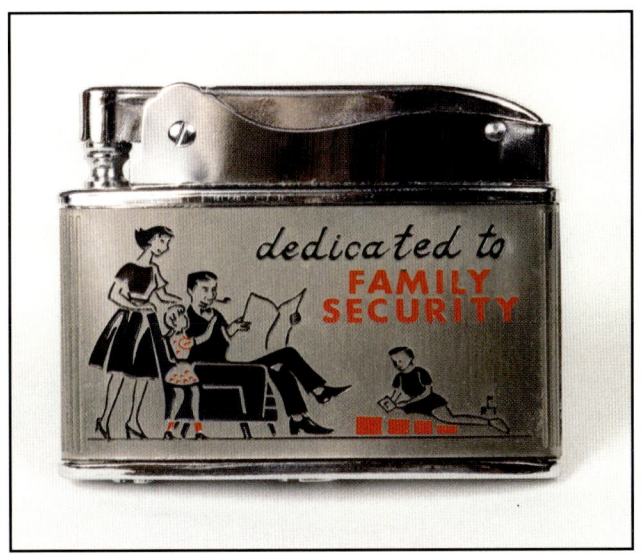

WELLINGTON-BALBOA, ca. 1955. Japanese flat advertising lighter for an insurance company. It reads "Dedicated to Family Security." $25-35.

Occupied Japan

Unknown, ca. 1948-1952. Two Occupied Japan piano shaped lighters. Both operate by pushing down on the keyboard. 3" tall. $50-90 each.

Occupied Japan, ca. 1948-1952. A rare, small version of the Leader radio. This one does not dispense cigarettes as the larger one does, but it does have a perpetual calendar. The right knob controls the lighter and the left, the calendar, with the small dial setting the month. $600-900.

Occupied Japan, ca. 1948-1952. The "Rocket" is activated by a turn of the propeller. Two other versions also exist. In one the two, fins are squeezed together to produce a light, and in the other there is no propeller but rather a "nose" which, when pushed in, opens the cockpit with a light. 4.5" l. $175-225.

Occupied Japan, ca. 1948-1952. Table lighter. Very unusual in that it is a "gas" occupied Japan lighter which is seldom seen and only one known, to date. System based on the French "Flamminaire." Shown with its original box. $250-350.

Occupied Japan, ca. 1948-1952. Made in Occupied Japan. A cowboy hat semi-automatic lighter. 4" l. $25-40.

Occupied Japan, ca. 1948-1952. This is an elegant little Lite Phone Microphone lighter with base. Chrome with white and black plastic. $175-225.

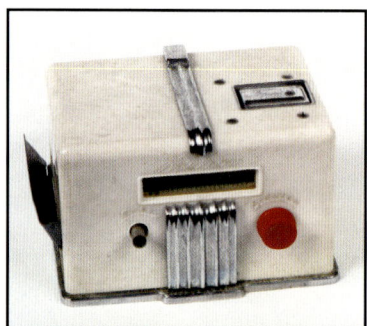

Occupied Japan, ca. 1948-1952. A plastic table lighter in the form of a bakelite radio. Made in Occupied Japan. It has a surprisingly complex mechanism. It could be loaded with twenty cigarettes. The left knob released a cigarette from the box. The dial recorded the number of cigarettes remaining. The right knob released the semi-automatic lighter action hidden in the top of the radio. 3" deep x 4.6" wide x 3.5" tall. $500-750.

Occupied Japan, ca. 1948-1952. A small fan shaped lighter. Turning the blade releases the snuffer. Other versions with wire cages around the blades were also produced. One of the many great figural lighters produced in Japan during the American occupation. Chrome plated brass. 4" h. $200-250.

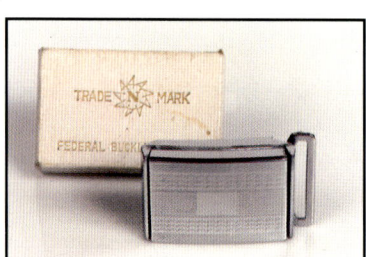

Occupied Japan, ca. 1948-1952. An unusual belt buckle lighter shown in open and closed positions. The lighter was removed from the belt buckle back to use. Produced in two different sizes as well as in a two-tone finish. 2.25" l. $50-100.

Miscellaneous

Chuckmuck, ca. 1880. An ornate brass form over a typical purse-like leather bag that would be used to hold the flint stone and tinder. It has a steel edge for striking the flint. It certainly looks like it was made in the Orient. 2.5" x 2.7". $150-200.

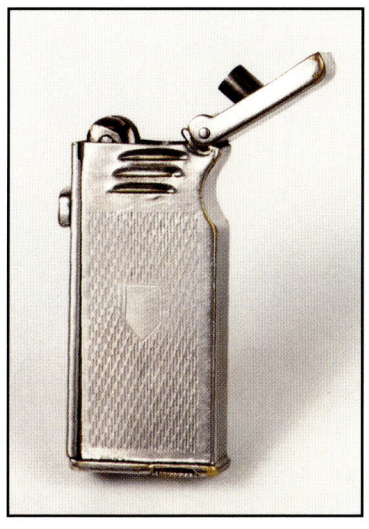

CELCO, Australia ca. 1942. An unusual, semi-automatic lighter with a built in windguard. It is quite unusual in that the entire flint assembly is raised as the mechanism is activated. Also manufactured in a manual version. Engine-turned brass with nickel plating. Marked "Upman Pat." and "Sydney." 2.7" x 1.2". $200-250.

FABERGE, ca. 1906. A Russian match safe with tinder cord and cigarette box in sterling silver with a gold washed interior. Provenance – Sotheby's Auction House. It has a cabochon clasp. 2.4" x 3.8". $3500-5000.

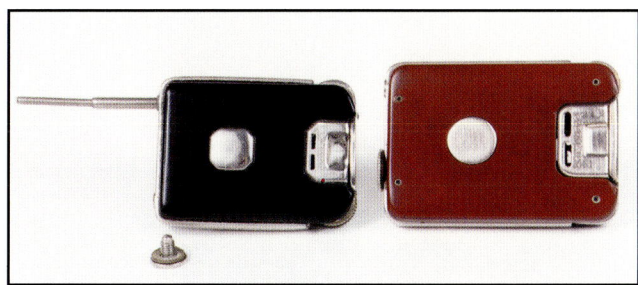

EVERMATCH, ca. 1933. Sweden. Here we see two versions of this lighter. The lighting mechanisms are the same – lift the cap and spin the triple wheel flint mechanism. The filing mechanism is unique – unscrew the fuel cap, lower the tube into the fluid, and pump the button. Rumor has it that one could also use the pump when windy, to force a fresh spate of fuel into the wick. Both are marked "Evermatch" on each side of the windguard. The smaller one (on the left) is marked on the side "MADE IN SWEDEN" followed by a long list of patent numbers. The larger version (on the right.) is marked on the side as well "PATENT PENDING" and "MADE IN SWEDEN." The larger one is plastic and brass, while the small one is enamel and brass. Also produced on a stand and resembling a microphone. $350-550.

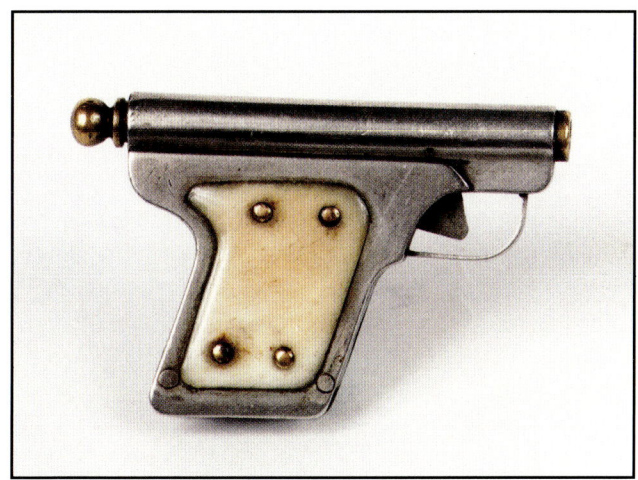

Unknown, ca. 1910. A figural pocket pistol (in the form of a semi-automatic gun) striker. Nickel plated brass with Ivorine handles. The striker material is on the bottom of the gun butt. 1.5" x 2.3". $300-400.

Unknown, ca. 1919. Nicely made but difficult to open figural book lighter with windguard. Remove the flint tube both to replace the flint and to refuel the lighter. No markings except as shown. Nickel plated brass. $100-150.

Unknown, ca. 1939. A chrome plated table lighter with an unusual clock shape. It is a heavy piece with a powerful spring loaded, semi-automatic action. 4.7" x 3.2". $250-300.

Unknown, ca. 1924. Celluloid wrapped, lift arm lighter with a built-in cigar cutter. Chrome plated brass. $300-350.

Unknown, ca. 1940. Very unusual squeeze mechanism, which activates the trap doors at the top, lifts the small wick cap, and activates the flint wheel in the proper order to achieve a light every time. Note the well designed windscreen. This lighter is chromium plated, completely unmarked and perhaps a prototype. $150-250.

Unknown, ca. 1934. A matchbook shaped, semi-automatic brass lighter with concealed mechanism. 2.2" x 1.2". $150-200.

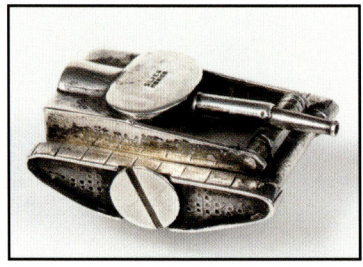

Unknown, ca. 1948. An exceptionally nice, small Mexican made silver lighter in the shape of a tank. The gun contains the spring for the flint and the mechanism is under the rear top plate of the lighter. Shown both open and closed. 2" l. $1,500-1,750.

Glossary of Terms and Abbreviations

COMPILED BY ALEXANDRE CLAPPIER

A. Pat. - Abbreviation for Auslands-Patent (German for "Foreign Patent").

Abalone - The shell lining of a gastropod mollusk, made of iridescent *mother-of-pearl*. Used in jewelry and incidentally in lighter ornamentation. Abalone is of a darker color than *mother-of-pearl*.

Acrylic - Artificial fiber made from vinyl compounds. Often used in the making of *Lucite* and *see-thru* lighters.

Alpaca - Alloy of *copper*, nickel and zinc. Also called Alpaka, Alpacca, German Silver, Nickel Silver, EPNS, Neusilber (German), Argentan (French), and mostly seen on German lighters.

Alpaka - See *Alpaca*.

Aluminum - The most abundant metal in the earth's crust, aluminum is a silver-white metal that's ductile, malleable, and an excellent conductor of heat and electricity. Aluminum was used mostly in Europe during and after World War II in the manufacturing of inexpensive and/or hand-made petrol lighters. Most of the hand-made lighters used aluminum recuperated from war surplus. Also called aluminum.

Amadou - French for *"tinder"*.

Angem. - Abbreviation for Angemeldet Patent (German for "Registered Patent").

Anodized (aluminum) - Metallic surface electrolytically coated with a protective or decorative oxide. Mostly used on *aluminum*.

Appliqué - A decoration or ornament made from a piece of one material applied to the surface of another. Mostly used on lighters as an *initial area* or ornament.

Argentan - French for *"alpaca"*.

Argenté (métal) - French for "silver-plated metal".

Art Deco - Derived from the 1925 Exposition Internationale des Arts Décoratifs et Industriels Modernes in Paris. A style of design popular during the 1920s and 1930s, and characterized by geometric and curved patterns, intense colors, and by the use of *hard glass enamel*, *lacquer*, exotic leathers, glass and porcelain among others.

Art Nouveau - A style of decorative art that originated from Western Europe and that was active in the late 19th and early 20th centuries. An ornamental and asymmetrical style characterized particularly by the depiction of plant tendrils, and exotic and symbolist themes. This style is often seen on early *striker* lighters, vesta cases, cigarette cases and tobacconia items decorated with *enamel*, *hard glass enamel*, and embossed silver.

Auermetal - Original name of the *synthetic flint* as invented by the Austrian scientist Auer von Welsbach in 1903, and made of an alloy of iron and *cerium*. Still in use today.

Automatic - As opposed to a *semi-automatic lighter*, a lighter that requires one motion to light and extinguish the flame. Ronson first introduced the concept of the automatic lighter circa 1926 with the Banjo.

Bakelite - A synthetic, thermosetting and fire-resistant resin invented in 1909 by L. H. Baekeland, and used as a substitute for *celluloid* and amber. Bakelite was used in the making of *pocket* and *table* lighters essentially in the 1930s and 1940s.

Basse taille - Engine turned surface covered with *hard glass enamel*.

Benzin - German for *"lighter fluid"*.

Brass - A ductile, yellow alloy of copper and zinc. Most often seen in *Trench Art* lighters.

Brevet Déposé - French for *"Patent Pending"*.

Breveté - French abbreviation for "Patent". Also seen as Bté.

Briquet - French for *"lighter"*.

Bronze - A yellowish and olive brown alloy of *copper*, tin, zinc, phosphorus and sometimes small amounts of other elements such as antimony. Bronze was used mostly in the making of large, ornamental lighters.

Brushed finish - As opposed to *high polish*, a surface that has been brushed to produce a non-reflective effect.

BSGDG - Stands for "Breveté Sans Garantie Du Gouvernement" (French for "Patent Without Warranty from the Government"). Also see *SGDG*.

Butane - An hydrocarbon gas obtained from natural gas or by refining petroleum..

Cap lighter - Used before the invention of the *synthetic flint*, a type of lighter using fulminate or mercury percussion caps, similar to the ones used in caps guns, to create the spark that ignites the *wick*.

Carat - French for *"karat"*.

Catalytic lighter - A category of lighter using a chemical reaction between denatured alcohol, air and platinum to create the spark that ignites the *wick*.

Celluloid - A synthetic plastic made from nitrocellulose and camphor. Celluloid was mostly used as a substitute for expensive substances such as ivory or tortoise shell, and is often found on lighters from the 1920s to the 1960s.

Cerium - A rare earth element used in the making of the *synthetic flint*. Also see *Auermetal*.

Chemical lighter - A type of lighter using a chemical reaction to create the spark that would ignite the fuel-soaked *wick*. Chemical lighters were common before the invention of the *flint* in the late 19th century and mostly in Europe. *Catalytic lighter* and *Potassium-bichro-*

mate lighters are examples of chemical lighters.

Chuck-muck - An early lighter originating from Tibet, the chuck-muck is considered as the ancestor of the *pocket lighter*. The chuck-muck is a *flint* and *tinder* lighter in the shape of a leather pouch to which a steel blade is attached.

Cloisonné - A decorative enamel work that originated from the Middle East. Cloisonné is made of metal filaments fused to the surface of an object which design is then filled with *enamel* of various colors.

Combination - A category of lighters that incorporates one or more different elements such as a compact or a cigarette case. Also called a combo.

Copper - A reddish metal whose characteristics are malleability, ductility and high conductivity. Because copper-made objects tend to form a patina when exposed to moist air, copper made lighters are usually plated with another metal such as silver, chrome or nickel.

Counter top lighter - A category of lighters for public use, often found on the countertops of bars and stores.

D.B.P. - Abbreviation for "Deutsches Bundes Patent" (German for "Patent of the German Federal Republic"). This marking can be found on German lighters made after 1949.

D.R.G.M. - Abbreviation for "Deutsches Reichs-Gebrauchs-Muster" (German for "German Reich Registered Patent"). This marking can be found on German lighters made between 1918 and 1945.

D.R.P. - Abbreviation for "Deutsches Reichs Patent" (German for "German Reich Patent"). This marking can be found on German lighters made between 1918 and 1945.

D.R.W.Z. - Abbreviation for "Deutsches Reichs Waren-Zeichen" (German for "German Reich Trade Mark").

De oro - Spanish for "made of gold".

Denatured alcohol - An ethyl alcohol usually mixed with *methanol* and used as fuel in certain lighters.

Déposé - See *Brevet déposé*.

Disposable lighter - A lighter meant to be disposed of when the tank is empty.

Dobereiner - Named after Johann Wolfgang Döbereiner (1780–1849), a German chemist who invented a lighting device in which hydrogen burns on contact with a platinum element.

Double-tank lighter - A lighter that incorporates a second, often detachable, fuel tank. The second tank was meant to provide a safety reserve of fuel.

Double-wheel - As opposed to *single wheel*, a lighter in which the *spark wheel* and the *roller* are two separate elements.

Dresden enamel - An ornamentation associating *basse taille* and hand painted motives, essentially found on Evans lighters.

Dureum - A gold-plated metal.

E.P.N.S. - Abbreviation for "Electro Plated Nickel Silver".

Electric lighter - A category of lighter in which the spark is created by electricity.

Enamel - Enamel is generally used to define all types of paint applied on a lighter such as enamel-based paint or *lacquer*. Not to be confused with *hard glass enamel*.

Engine turning - As opposed to hand engraving, a type of engraving made by machine. Also called machine engraving.

Feuerzeug - German for "lighter".

Figural lighter - See *Novelty lighter*.

Filler screw - Screw used to hermetically close the fuel tank of a lighter.

Flat advertiser - A category of lighters bearing advertising. Most of them were made in Japan and in the shape of the Ronson Adonis.

Flint - A very hard type of quartz whose characteristic is to create sparks when struck with steel, flint was used in primitive lighters such as the *chuck-muck* or the Japanese tinder lighters (see *Japanese walnut*). Flint is also used to define the synthetic flint made of an alloy of *cerium* and iron. Also see *Auermetal*.

Flint screw - The screw that keeps the flint spring in place and under pressure, so that the *flint* is pressed against the *spark wheel*.

Flint tube - The element in a lighter that contains the *flint*.

Fluid - Liquid fuel made of petroleum distillate. Usually also called petrol or fuel.

For. Pat. - Abbreviation for "Foreign Patent".

Frozen flint - Flints disintegrate over time and tend to block the *flint tube* and the *spark wheel*. Under the pressure of the flint spring, the flint eventually gets stuck in the *flint tube*, hence the term frozen flint.

Fuel - Generic terms for liquid fuels made from petroleum, such a kerosene, gasoline or *lighter fluid*.

Fusee lighter - See *Tinder lighter*.

Galuchat - French for "shagreen".

Gas - Generic term for *butane*. Gas also can mean petroleum fuel in the United Kingdom.

German silver - See *Alpaca*.

Ges. Gesch. - Abbreviation for "Gesetzlich Geschutzt" (German for "Registered").

Gilded - Said of a surface covered with a thin layer of *gold*.

Glass enamel - See *Hard glass enamel*.

Gold - Pure gold is a very ductile and malleable precious metal. Gold is usually hardened by alloying pure gold with other metals such as copper, nickel or silver. The content of pure gold in a gold alloy is stated in *karats* or sometimes in terms of parts per thousand. As an example, 18-karat gold contains 75% of pure gold, and could be also expressed as a gold with a fineness of 750.

Guilloché - French for *engine turning*.

Hallmarks - Originally used in England, a mark found on silver and *gold* to signify a certain level of purity. A hallmark is often accompanied by various marks stating the date when the piece was manufactured, the name of the maker, and the place where the piece was actually hallmarked.

Hand-chased engraving - As opposed to *machine engraving* or *engine turning*, an engraving made by hand.

Hard glass enamel - A decorative vitreous coating baked on metal. Usually transparent, opaque and with or without color, hard glass enamel is mostly used in the *cloisonné* and *bassse taille* enamelware.

High polish finish - As opposed to *brushed finish*, a surface that has been polished to produce a reflective effect. Also called high gloss finish.

Initial area - See *Monogram plaque*.

Japanese walnut - A netsuke and an early lighter using *tinder* and *flint*, and made at the size of a walnut.

Jump spark lighter - A category of *electric lighters* where ignition of the *wick* is created by arcing electricity.

Karat - Karat is equal to 1/24 part and is a unit of measurement for the fineness of gold. 24-karat gold is pure gold while 12-karat gold is an alloy made of 50% pure gold and 50% of another element. The most common fineness of gold used in the making of lighters are 9, 14 and 18-karat. Abbreviations of Karat are K, Kt., C and Ct. Also see *Carat*.

Lacquer - A natural or synthetic solution, lacquer is usually applied as an ornamental or protective coating to give a high gloss finish to surfaces. Also see *Maki-e* and *Namiki*.

Leather cover - Lighter whose body is wrapped in leather. Also called leather wrap.

Lift-arm - A category of *manual* lighters where the *snuffer* is attached to an arm. The first lift-arm lighter was marketed by Dunhill in 1922 with the Everytime Lighter, soon to become the Dunhill Unique.

Lighter - By the time you read this entry, you should pretty much know what a lighter is. If you do not, a lighter is a device used to light a cigarette, a cigar or a pipe, and more generally any source of combustion.

Lucite lighter - Lucite is a transparent thermoplastic acrylic resin. Also see *See-thru lighter*.

Machine engraving - See *Engine turning*.

Made in Occupied Japan - A category of lighters made in Japan during the US occupation, from 1945 to 1952, and stamped accordingly. Abbreviation is MIOJ.

Maki-e - A technique of Japanese *lacquer* that incorporates designs made out of colored pigments, silver and gold powder, and inlaid materials such as *mother-of-pearl*. There are four forms of Maki-e lacquer: Hira Maki-e (flat), Togidashi Maki-e (burnished), Taka Maki-e (relief) and Raden Maki-e (inlay). Maki-e lacquer is most often seen on Dunhill lighters, called Dunhill *Namiki*.

Manual - As opposed to *semi-automatic* and *automatic* lighters, a lighter that requires more than two motions to ignite and extinguish the flame.

Methanol - A colorless and flammable liquid used as fuel in certain lighters. Also called wood alcohol, wood spirits and Spirit of Columbia.

MIOJ - See *Made in Occupied Japan*.

Modernistic - Another name for the *Art Deco* style and used at the time when the term Art Deco was not in use yet.

Monogram plate - A metal *applique* whose purpose is to be engraved and personalized with the owner's initials or monogram. Also called initial plate.

MOP - See *Mother-of-pearl*.

Mother-of-pearl - Also called *nacre*, the pearly substance that forms the lining of certain mollusk shells. Mother-of-pearl is often used as an ornamentation on lighters. Also see *abalone*.

Namiki - The type of *Maki-e lacquer* found on Dunhill lighters. Named after the Namiki Manufacturing Company which was in charge of decorating Dunhill lighters and pens with Maki-e lacquer.

Nickel silver - See *Alpaca*.

Novelty lighter - A category of lighters made in the shape of typical and figural objects. Also called *figural lighter*.

Pat. Ang. - Abbreviation for "Patent Angemeldet" (German for "Patent Pending").

Patent Applied For - See *Patent Pending*.

Patent Pending - Often found on lighters, a term which is meant to inform the public that an application for patent has been filed with the Patent and Trademark Office. It has the same meaning as Patent Applied For.

Pearloid - Similar to *celluloid*, a synthetic plastic made to look like *mother-of-pearl*.

Percussion cap - See *Cap lighter*.

Pocket lighter - As opposed to *table lighter*, a lighter designed to be carried in a pocket.

Potassium-bichromate lighter - An early *chemical lighter* using zinc and carbon to create the *spark* that ignites the *wick*.

Primitive lighter - A lighter using elements found in nature such as *flint*, wood or *tinder*. Also see *Chuck-muck*.

R.P. - Abbreviation for "Reichs Patent" (German for "Reich Patent"). This marking can be found on German lighters made between 1871 and 1918. Also see *D.R.G.M.*

Roller - A knurled piece of metal of various length used to activate the *spark wheel*.

See-thru lighter - A category of lighters in which the body is made of objects encased in a translucent resin such as *Lucite* or *acrylic*.

Semi-automatic - As opposed to *automatic*, a lighter that requires two separate motions to light and to extinguish the flame.

SGDG - Stands for "Sans Garantie Du Gouvernement" (French for "Without Warranty from the Government"). A term meant to inform the public that the design of a lighter has been duly patented. The SGDG marking can be found on French lighters manufactured between 1844 and 1968. SGDG has the same meaning than *BSGDG*.

Shagreen - The hide of a shark or ray used as an ornamental leather cover on lighters. Shagreen is also used to describe grainy leather usually dyed green. Also see *Galuchat* and *Sharkskin*.

Sharkskin - Leather made from the hide of a shark. Also see *Shagreen*.

Silent flame - A category of *electric lighters* that use batteries to heat a coil which ignites the *wick*. Silent flame lighters were made mostly by Parker and Dunhill.

Single wheel - As opposed to *double wheel*, a lighter on which the *spark* wheel must be operated directly without the help of a second wheel or *roller*.

Snuffer - Originally the instrument used to snuff out a candle. On a lighter, the snuffer is the part that covers the flame to extinguish it.

Spark wheel - The toothed wheel that creates sparks by rotating and rubbing the *flint*.

Spirit of Columbia - See *Methanol*.

Sterling silver - An alloy containing a minimum of 92.5% pure silver. Lighters made of sterling silver usually bear a marking such as Sterling, 925, or a *hallmark*.

Strike lighter - See *Striker*.

Striker - A lighter with a wand that combines a fuel-soaked *wick* and a steel tip to be scratched along a strip of *flint*. Also called strike lighter, metal match and scratcher.

Synthetic flint - Another name for the flint made of an alloy of *cerium* and iron. Also see *Auermetal*, *Cerium* and *Flint*.

Table lighter - As opposed to *pocket lighter*, a lighter designed for home use or of a size too large to be carried in a pocket.

Tax seal - A seal soldered to a lighter and which served as proof of the payment of the tax on lighters. Tax seals were present on lighters sold in France and Belgium in the first half of the 20th century.

Thumb lever - Lever that activates the lighter, meant to be operated with a thumb.

Thumb wheel - Spark wheel meant to be operated with a thumb. Also see *Single wheel*.

Tinder lighter - Lighter using a *tinder* as a *wick*, which smolders instead of burns when ignited. Tinder lighters are known to be highly *windproof*. Also called *woven tinder lighter* or *fusee lighter*.

Trench Art lighter - Category of lighter made by soldiers in trenches during World War I. Most lighters were made of leftovers and parts found in the trenches such as buttons, coins and cartridge cases. The term is also used for lighters made by veterans after the war was over.

Wand - A metal rod that contains a fuel-soaked *wick*, and used in a *striker*.

Watch lighter - A lighter that incorporates a watch.

Wick - A cord or strand of fibers that draws up fuel to the flame by capillary action.

Windguard - See *Windscreen*.

Windproof - Category of lighters that are resistant to wind. The most famous example of a windproof lighter is the Zippo.

Windscreen - Piece of metal that surrounds the flame in order to protect it from the wind.

Woven tinder - Another name for *tinder*.

Woven tinder lighter - See *Tinder lighter*.

Zamac - Alloy of zinc and *aluminum* often used in the making of low-quality lighters. Zamac stands for "Zinc Aluminum Metal Alloy Casting."

Zunder - German for "*tinder*."

Bibliography

Clayton, Larry, *The Evans Book*, 1998, Schiffer Publishing, Atglen, PA.

Bisconcini, Stephano, *Lighters/Accendini*, 1983, Edizioni San Gottardo, Milano, Italy.

Cummings, U.K., *Ronson - The World's Greatest Lighter*, 1992, Bird Dog Books, Palo Alto, CA.

Schneider, Stuart and Ira Pilossof, *Handbook of Vintage Cigarette Lighters*, 1999, Schiffer Publishing, Atglen, PA.

Schneider, Stuart, and George Fischler, *Cigarette Lighters*, 1996, Schiffer Publishing, Atglen, PA.

Van Weert, Ad and Joop Bromet, *The Legend of the Lighter*, 1995, Abbeville Press, NY

Resources

Authorized Repair Service, 30 W. 57th Street, New York, New York 10019. Dealers in vintage lighters and does lighter repair.

International Lighter Collectors Club, c/o Judith Sanders, P.O. Box 1733, Quitman, Texas 75783. The largest lighter club. Publishes "OTLS" *On The Lighter Side* newsletter. Each issue of the newsletter has information on the lighter shows, articles about lighter companies, classified advertising, etc.

Pocket Lighter Preservation Guild, c/o Larry Marshall. P.L.P.G., P.O. Box 327, Wentzville, MO 63385-0327, PH: 314-651-0693, FAX: 636-639-8990. Lighter club.

Vintage Lighters, Inc., Ira Pilossof P.O. Box 1325, Fair Lawn, New Jersey 07410. Dealers in vintage lighters. Phone 201-797-6595 E-mail vintageltr@aol.com